THE U.S. COMBAT AND TACTICAL WHEELED VEHICLE FLEETS
ISSUES AND SUGGESTIONS FOR CONGRESS

TERRENCE K. KELLY • JOHN E. PETERS
ERIC LANDREE • LOUIS R. MOORE
RANDALL STEEB • AARON MARTIN

Prepared for the Office of the Secretary of Defense

RAND NATIONAL DEFENSE RESEARCH INSTITUTE

The research described in this report was prepared for the Office of the Secretary of Defense (OSD). The research was conducted within the RAND National Defense Research Institute, a federally funded research and development center sponsored by OSD, the Joint Staff, the Unified Combatant Commands, the Navy, the Marine Corps, the defense agencies, and the defense Intelligence Community under Contract W74V8H-06-C-0002.

Library of Congress Control Number: 2011921602

ISBN: 978-0-8330-5173-8

Cover design by Pete Soriano

Published 2011 by the RAND Corporation
1776 Main Street, P.O. Box 2138, Santa Monica, CA 90407-2138
1200 South Hayes Street, Arlington, VA 22202-5050
4570 Fifth Avenue, Suite 600, Pittsburgh, PA 15213-2665
RAND URL: http://www.rand.org/
To order RAND documents or to obtain additional information, contact
Distribution Services: Telephone: (310) 451-7002;
Fax: (310) 451-6915; Email: order@rand.org

Preface

This monograph considers the challenges confronting the military services, principally the U.S. Army and the U.S. Marine Corps, in fielding the ground combat vehicle fleet and armored tactical wheeled vehicle fleet of the U.S. Department of Defense (DoD). In Section 222 of the National Defense Authorization Act for Fiscal Year 2010 (Pub. L. 111-84), Congress mandated the study that produced this monograph, specifying that the study would provide the following:

(A) A detailed discussion of the requirements and capability needs identified or proposed for current and prospective combat vehicles and armored tactical wheeled vehicles.

(B) An identification of capability gaps for combat vehicles and armored tactical wheeled vehicles based on lessons learned from recent conflicts and an assessment of emerging threats.

(C) An identification of the critical technology elements or integration risks associated with particular categories of combat vehicles and armored tactical wheeled vehicles, and with particular missions of such vehicles.

(D) Recommendations with respect to actions that could be taken to develop and deploy, during the ten-year period beginning on the date of the submittal of the report, critical technology capabilities to address the capability gaps identified pursuant to subparagraph (B), including an identification of high-priority science and technology, research and engineering, and prototyping opportunities.

This line of inquiry is consistent with the sentiment expressed in the February 2010 Quadrennial Defense Review, which observed,

> Another pressing institutional challenge facing the Department is acquisitions—broadly speaking, how we acquire goods and services and manage the taxpayers' money. Today, the Department's obligation to defend and advance America's national interests by, in part, exercising prudent financial stewardship continues to be encumbered by a small set of expensive weapons with unrealistic requirements, cost and schedule overruns, and unacceptable performance.[1]

This monograph is a response to a congressional request. However, it will also be of interest to U.S. military personnel involved in the acquisition and management of vehicles and their attendant systems, as well as to contractors and manufacturers that furnish the systems and technologies discussed here.

This research was sponsored by the Director, Defense Research and Engineering (DDR&E), Office of the Secretary of Defense, and conducted within the Acquisition and Technology Policy Center of the RAND National Defense Research Institute, a federally funded research and development center sponsored by the Office of the Secretary of Defense, the Joint Staff, the Unified Combatant Commands, the Navy, the Marine Corps, the defense agencies, and the defense Intelligence Community.

For more information on the RAND Acquisition and Technology Policy Center, see http://www.rand.org/nsrd/about/atp.html or contact the director (contact information is provided on the web page).

---

[1]   U.S. Department of Defense, *Quadrennial Defense Review Report*, Washington, D.C., February 2010, pp. 75–76.

Contents

Figures

Tables

Summary

In Section 222 of the National Defense Authorization Act for Fiscal Year 2010 (Pub. L. 111-84), Congress directed the Secretary of Defense to contract with an independent body to assess activities for technology modernization of the ground combat vehicle and armored tactical wheeled vehicle fleets, and specifically to

- provide a detailed discussion of requirements and capability **needs**
- identify capability **gaps** for vehicles based on lessons learned from recent conflicts and an assessment of emerging threats
- identify critical technology elements or integration **risks** associated with particular categories of vehicles and with particular missions of such vehicles
- recommend **actions** to develop and deploy critical technology capabilities to address the identified capability gaps.[1]

Methods of Inquiry

This monograph reflects the results of the ensuing effort. Figure S.1 illustrates the basic process and the way in which the project's tasks culminated in a series of recommendations.

As the figure suggests, the research was organized around four tasks. The first task examined the requirements and capability needs

---

[1] Bold text highlights the logical links among needs, gaps in meeting them, risks that attend enduring gaps, and actions that seek to close them.

Figure S.1
Research Process and Recommendations

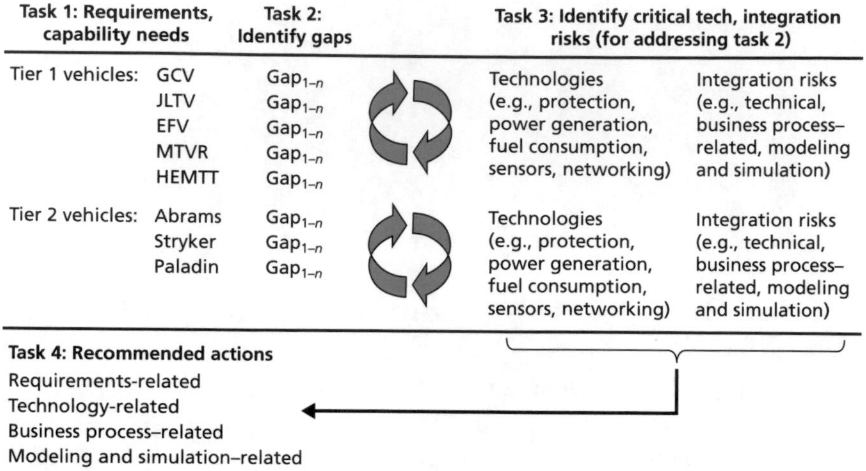

Task 1: Requirements, capability needs		Task 2: Identify gaps	Task 3: Identify critical tech, integration risks (for addressing task 2)	
Tier 1 vehicles:	GCV	Gap$_{1-n}$	Technologies (e.g., protection, power generation, fuel consumption, sensors, networking)	Integration risks (e.g., technical, business process–related, modeling and simulation)
	JLTV	Gap$_{1-n}$		
	EFV	Gap$_{1-n}$		
	MTVR	Gap$_{1-n}$		
	HEMTT	Gap$_{1-n}$		
Tier 2 vehicles:	Abrams	Gap$_{1-n}$	Technologies (e.g., protection, power generation, fuel consumption, sensors, networking)	Integration risks (e.g., technical, business process–related, modeling and simulation)
	Stryker	Gap$_{1-n}$		
	Paladin	Gap$_{1-n}$		

Task 4: Recommended actions
Requirements-related
Technology-related
Business process–related
Modeling and simulation–related

NOTE: GCV = ground combat vehicle. JLTV = joint light tactical vehicle. EFV = expeditionary fighting vehicle. MTVR = medium tactical vehicle replacement. HEMTT = Heavy Expanded Mobility Tactical Truck.
RAND MG1093-S.1

for a cluster of vehicles selected in cooperation with the sponsor to reflect attributes that might typify the universe of such vehicles. The research team collected official requirements and design documents from the Army and Marine Corps organizations responsible for them. Task two, identifying capability gaps, involved reviewing the official documentation for each vehicle, talking with Army and Marine Corps officials involved in the vehicles' development and fielding, and arriving at a list of capability gaps for the vehicles in the sample.

Once the capability gaps were identified, the research effort, through task three, identified critical technologies and integration risks for addressing the gaps in question. As the figure indicates, the research team was able to identify technology domains (i.e., protection, power generation, fuels and fuel consumption, sensors and networking) in which the most important capability gaps reside and the integration risks attending them. These risks lay in the technologies identified to close the capability gaps (e.g., immature, high-risk), business processes used by the Army and Marine Corps in managing the initiatives producing these vehicles or supervising their modifications and recapitalization, and modeling and simulation (M&S) in support of the research, development, and acquisition efforts that brought each vehicle into being.

At the point where the research team had formulated tentative observations about the issues confronting combat and tactical wheeled vehicle's research, development, and acquisition, we held a workshop to vet our findings with the DDR&E; the Office of the Under Secretary of Defense for Acquisition, Technology, and Logistics and other stakeholders in the Office of the Secretary of Defense; and Army and Marine Corps stakeholders. The workshop featured a plenary session in which the project team briefed its observations and some tentative findings and subsequently lead smaller "breakout" sessions to capture the reactions of subject-matter experts. There were three such breakout groups: one for technology, one devoted to business processes, and one for M&S. These sessions became the engine that drove the research effort to task four, identifying recommended actions.

Observations on Closing Capability Gaps and Meeting Performance Requirements

Closing capability gaps and addressing performance requirements are difficult tasks. Part of the difficulty arises from the cycle of action-reaction between U.S. and enemy forces as they seek tactical advantages over each other. With tactical wheeled vehicles like the HEMTT, part of the difficulty lies in the fact that these vehicles are "repurposed" commercial trucks.[2] The more extensive the modifications necessary to close capability gaps or satisfy current performance expectations, the more expensive the work is likely to be. The challenge, however, is not limited to tactical wheeled vehicles.

Vehicles must provide the ability to manage the competing demands of operational requirements for power, protection, and pay-

---

[2] An observation from an engineer and branch chief at U.S. Army Training and Doctrine Command (TRADOC) during discussions with the project leaders, October 20, 2010.

load or performance that induce size, weight, power, and cooling (SWaP-C) requirements. Sometimes, one of these considerations (e.g., protection) dominates the equation and thus dominates the other design criteria, as shown later in our discussion of the GCV. Sometimes, the appearance of new operational requirements can cause a new vehicle to evolve as a much different and more expensive vehicle than the one it replaces, as illustrated by the case of the JLTV, the successor to the high-mobility multipurpose wheeled vehicle (HMMWV). The current emphasis on affordability places additional constraints on the services' ability to manage SWaP-C requirements, develop materiel solutions to close capability gaps, and satisfy evolving performance requirements, as the Paladin Integrated Management program case demonstrates.[3] There are also instances in which a vehicle can play such a central role in fulfilling a service's mission, as the EFV does in the Marine Corps, that the service will accept lengthy schedule delays, significant cost growth, and substantially revised performance criteria to preserve the capabilities that the vehicle is expected to provide.[4] Finally, Stryker illustrates what can happen when circumstances surrounding a vehicle change dramatically. The Stryker began life as an interim vehicle until the Future Combat Systems (FCS) reached fruition and emphasized air-transportability for crisis responsiveness over other design considerations. Subsequently, the FCS was canceled, thrusting the Stryker into a new, extended role as a major part of Army force structure under circumstances substantially different from those anticipated for the vehicle when it was being acquired initially. As a result, it continues to have capability gaps in protection, mobility, lethality, and networking, despite vigorous Army efforts to adapt the vehicle for current opera-

---

[3] Affordability has always been a concern, but Under Secretary of Defense Ashton Carter recently emphasized this issue in a memorandum for the secretaries of the armed services and directors of defense agencies (Ashton B. Carter, Under Secretary of Defense for Acquisition, Technology, and Logistics, "Better Buying Power: Guidance for Obtaining Greater Efficiency and Productivity in Defense Spending," memorandum to acquisition professionals, September 14, 2010c).

[4] As of this writing, Secretary of Defense Robert Gates has announced the cancellation of the EFV, but its fate remains undetermined as its proponents seek to challenge the decision.

tions, and illustrates the challenges of forecasting requirements far into the future.

Modeling and Simulation

M&S can be helpful in this regard by identifying the limits to on-vehicle trade-offs and representing possible off-vehicle effects. Operational M&S can help with the problem raised earlier and can assist the combat developer and acquisition communities in defining operational and program requirements in the first place; program-oriented M&S designed to inexpensively evaluate individual vehicles (or other systems) or system components is critically important.

The primary challenge to M&S in the U.S. Department of Defense (DoD) appears to be in maintaining the agility to keep up with changing battlefield requirements and ensuring the development of new tactics and technologies. Among the trends we have seen that require a more streamlined and efficient analysis process are the following:

- Nontraditional acquisition and modular electronic architectures require more agile M&S processes to keep up with changing requirements.
- The trade-off of overlapping protection devices means that M&S must be able to quickly represent both the new capability and the impact on space claims, mobility, and power.
- Models will need to better represent special functions for electronic warfare, communication jamming, and interoperability to provide better situational awareness.

Observations and Conclusions

Our analysis led to the following observations and conclusions.

Requirements-Related Issues

In our interactions with combat developers from the Army and Marine Corps, we found no evidence of fundamental flaws in their requirements development processes for the vehicles we considered. We were able to observe that arriving at a satisfactory set of requirements for tactical wheeled and ground combat vehicles is complicated by the fact that the vehicles remain in the services' inventories for decades. Combat developers typically have a deep understanding of current and near-term operating requirements, but they cannot unfailingly predict the future.

The implications of these circumstances are that, all but inevitably, *DoD will have vehicles in its fleets that were designed and built for requirements other than those it finds itself facing in the future.* This fact is driven by the wide spectrum of potential threats and scenarios in the 21st century and the fundamentally different physics and engineering problems presented by these threats. Choices will have to be made.

The full set of desired operational requirements is unlikely to be met in many cases. Because of the constraints on the trade space into which all vehicle requirements must fit, the resulting vehicles are unlikely to deliver 100-percent performance against all desired design criteria.

The iron triangle of trade-offs is permanent.[5] In particular, DoD will always want vehicles that provide better protection, have more power (electrical and mechanical), and perform better or are more capable (in terms of weight, mobility, and so on). No matter what technical advances are made, there will always be a drive to do better in these categories, and advances will help protect soldiers and marines; make the U.S. military more mobile strategically, operationally, and tactically; and increase performance. Investments in these areas will always be beneficial. As a result, *the vehicles resulting from this process may fail to meet all requirements but may nevertheless be satisfactory.*

These observations with respect to requirements have implications for technology- and engineering-related issues, as well as for acquisition-related processes. The technology- and engineering-related

---

[5] The "iron triangle" of trade-offs are those among performance (recently emphasizing power), protection, and payload.

issues most closely align with the questions asked by Congress; however, they will be much more likely to come to fruition at reasonable costs and within reasonable time frames if the acquisition process issues are also addressed.

Technology-Related Issues

The study found four major technology-related issues associated with the vehicle fleets to be the most challenging with respect to meeting operational requirements. They were protection, power generation, fuels and fuel consumption, and sensors, networking, and complexity. We treat each issue in turn.

Protection. The critical observations with respect to protection are as follows:

- Protection requirements differ based on expected threats, and technical and engineering solutions will differ based on these requirements.
- Protection requirements consider onboard and offboard technology as well as vehicle design and integration improvements.
- Improving protection will be a permanent task to which technology and engineering will need to contribute (along with tactics, unit designs, and other factors); it will never be good enough.

Electrical Power Generation. The critical observations with respect to power generation are as follows:

- The advent of tactical networks, computer-based battle command systems, and expectations of battle command on the move, situational awareness, and various protection devices drive demand for electrical power upward. This trend will likely continue.
- In some instances, fitting larger alternators onto the vehicles to supply the necessary power is adequate, but in other cases, it is not. Some vehicles require large battery storage, fuel cells, or auxiliary power units to provide the necessary electricity and associated capabilities.

- Reducing the need for external generators and associated equipment and support enhances strategic and operational mobility and reduces logistical requirements.
- The demand for additional electrical power means that vehicles must be able to not only provide the electricity but also accommodate the space, weight, and cooling requirements associated with the additional equipment. The vehicle's space, weight, power, and cooling capabilities must be flexible enough to accommodate new equipment that evolves later.
- Their designs must have "open architecture" to accommodate future network-related equipment, along with the additional weight and space this equipment will claim on the vehicle and the heat the new components will generate.
- The demand for additional electricity affects the designs of both tactical wheeled and ground combat vehicles.
- Future vehicles will almost certainly be more expensive than their predecessors, in part because they will need advanced power-generating capabilities.

Fuels and Fuel Consumption. The critical observations with respect to fuels and fuel consumption are as follows:

- Fuel costs and availability are major factors in ongoing and possible future operations.
- The fully burdened cost of fuel and the logistics requirements for supplying fuel on the battlefield are important and not always taken fully into consideration during acquisition.
- Future conflict could pose even more challenges with respect to fuel, such as if U.S. forces were unable to secure enough fuel from international supply routes, forcing them to depend on local fuels (which at the moment they cannot use in many places without damaging some equipment).

Sensors, Networking, and Complexity. Sensors and networks are outside the formal purview of this study, but due to their significant effects on many aspects of vehicle design and modification—

including protection, power, space, and cooling considerations and providing required vehicle capability—how these functions develop and are implemented must be briefly considered. They arc critical technologies that will be important considerations for Congress to examine. Given this fact, the critical observations with respect to sensors and networking are as follows:

- Sensors and networking contribute to vehicle complexity, which represents the possibility that unidentified dependencies and incompatibilities among components and subsystems will cause systems to fail.
- Hedging against the effects of complexity requires additional efforts in systems engineering and systems integration, with the understanding that some aspects of complexity are not well understood and thus cannot be easily identified and fixed.
- Complexity adds a greater chance for schedule slippage and cost growth for the vehicles currently under development than there was with their simpler predecessors.
- Increased complexity is the result of efforts to develop greater operational capabilities and better meet operational requirements. It cannot be done away with, so it must be well managed.

Acquisition Policy and Business Process–Related Issues

At least seven key observations based on prevailing DoD policies and business processes bear on the services' ability to field vehicles that are appropriate for the anticipated operating circumstances. These include the following:

- *The funding implications of the survivability of tactical wheeled vehicles:* As a result of current operations, tactical wheeled vehicles are acquiring more situational awareness and protection capabilities, thus growing closer to their ground combat vehicle cousins and more distant from their commercial counterparts. These trends mean more expensive vehicles in most fleets and, due to the large number of tactical wheeled vehicles, much more expensive fleets.

The trade-off between survivability and affordability presents a major policy decision for DoD and Congress.

- *Stable funding and vehicle requirements:* Many acquisition officials believe that funding instability and creeping vehicle requirements are among the biggest threats to their programs.[6]
- *Cost-estimating procedures:* Among the officials interviewed for this work who commented on cost estimating, most believed that life-cycle estimates were superior to unit cost estimates and that different acquisition decisions would be made and net life-cycle costs reduced if cost estimates more thoroughly included these considerations.[7]
- *Aligning the proper M&S tools to support decisions and decisionmakers:* M&S efforts do not appear to be fully aligned with the decisions they are meant to support (e.g., whether a materiel solution is warranted, technology development, analysis of alternatives, milestone decisions) and the information needs of the officials who will make them. If the services are to enjoy the full benefits of the M&S conducted to assist with the research, development, and support of vehicle programs, they must make a greater effort to perfect this alignment.
- *Acquisition category (ACAT) decisions that emphasize risk rather than just cost:* Risk, of which program cost is an important element, should be the dominant factor in ACAT decisions. Risk, the minimizing of which is the driving concept behind the decision, is not currently considered, except to the extent that cost is used as a proxy for risk. As a result, mature, well-understood, but expensive programs contemplating changes and modifications

---

[6] Funding can be unstable for a variety of reasons, including service or DoD funding decisions, changes in program costs that have the same effect as changes in funding levels (these two being the most common causes), and changes in congressional priorities (e.g., when Congress requires changes in programs that affect overall plans and budgets). This monograph does not examine these causes in detail but does note that unstable funding was the most frequently stated concern among program managers with whom the research team met.

[7] There are indications that some of these concerns may be addressed through pending changes to acquisition practices. See the directions on how to consider cost estimates in the memorandum from Under Secretary Carter (2010c).

that pose little risk are nevertheless subjected to stringent requirements meant to manage risk.

- *Adequately resourcing programs from the beginning:* The consensus among the experts with whom we spoke emphasized the need to ensure that programs are appropriately resourced from the outset. This is particularly important for large, complex programs for which having the right managerial and technical talent in place early on is essential for success. While there are real challenges to ensuring that this happens, it is a critical element in the success of complex systems.
- *More fully integrated test and evaluation:* A number of experts interviewed for this work noted that, in practice, independent tests and evaluations sometimes led to new performance requirements for vehicles emerging at the end of a system's development that were not represented in the requirements documents. This late appearance of new performance criteria sometimes led to delays in final certification for the vehicle and often added to program cost and caused schedule delays as the program tried to satisfy the newly evolved standards. Testing and evaluation activities that are more closely integrated throughout the program's development would be more helpful.

Trends

Equipping the armed services with tactical wheeled and ground combat vehicles will remain a challenging endeavor for the multitude of reasons cited throughout this monograph and summarized here. Some factors are clearly positive and should help ensure the acquisition of vehicles suitable for the anticipated circumstances. Some factors are clearly negative and complicate the task that the services face in equipping their forces. Still other factors are ambiguous at this point but could prove to be positive or negative as their effects become more visible.

Positive Trends

The preference among program managers for relatively mature technologies at the beginning of the technology development phase of vehicle programs is clearly positive. The practice reduces dependencies on immature technologies that can lead to cost growth and schedule slippage when they do not develop as quickly as estimated. The practice also increases the probability that the technologies that are central to the vehicle's success will be more fully developed than otherwise would be the case and will therefore avoid negatively affecting engineering and manufacturing development.

The services' appreciation of systems engineering expertise is another positive development. Both the Army and Marine Corps seem to recognize the centrality of systems engineering to program success and appear to be trying to grow their capacity in this field.

The services have renewed their efforts to improve management practices and risk management, typified by knowledge points, competitive prototyping, gate reviews, portfolio reviews, requirements-stabilization initiatives, and other efforts described in Chapters Two and Three. To be sure, there is room for improvement here—for example, by insisting that all programs satisfy the criteria for being "born healthy," as discussed in Chapter Three.

Another positive sign lies in the responsiveness of the research, development, and acquisition communities. Collectively, they have shown an improved ability to produce needed vehicles in a hurry: mine-resistant, ambush-protected (MRAP) vehicles and the MRAP all-terrain vehicle (M-ATV) are good examples. They have also demonstrated responsiveness to Joint Urgent Operational Needs Statements and Operational Needs Statements, fielding B-kit armor for HEMTTs and HMMWVs, among other responses.

Modeling, simulation, and experimentation also hold great potential for improving vehicle designs, especially if this field evolves along the lines suggested in Chapter Five.

Negative Trends

If necessity continues to drive tactical wheeled vehicle requirements closer to those of their ground combat vehicle cousins, that will surely have the salutary effect of affording their crews greater protection and situational awareness, but these positive developments will be accompanied by complexity and cost growth. As a result, new vehicles will almost certainly be significantly more expensive than the ones they replace.

This phenomenon will probably manifest across all vehicle fleets as recapitalization and replacement go forward. In addition, as the case of the JLTV suggests, there may be a divergence in requirements between the Marine Corps and the Army to meet performance criteria exclusive to each service. If this occurs, the unit costs of the vehicles in question will probably increase because of the loss of economies of scale when each service procures its own designs.

Also, there is the persistent vulnerability of the vehicle fleets to adaptive threats. As the GCV example suggests, this state of affairs can emphasize design criteria in favor of protection, and it can compromise all other performance dimensions in the process. Technology-based solutions to mitigate vulnerability are expensive, whereas the enemy's countermeasures are relatively cheap. It is impossible to protect the vehicle fleets from all threats solely with onboard armor, situational awareness, and active protection systems. At the same time, incorporating off-vehicle assets in trade-offs and calculations of vehicle requirements necessitates further assistance from the M&S community.

Uncertain Trends

The potential of robotics and autonomous systems, on its face, seems significant. Perhaps it is, but until the services advance these technologies and develop concepts for their application in roles that would reduce the threat to tactical wheeled and ground combat vehicles, their future utility remains unclear. Removing soldiers and marines from harm's way is an important but perhaps insufficient contribution, especially if the costs associated with the systems in question rival or exceed those of the manned vehicles they replace.

The effects of the network on vehicles are another question mark. The FCS program revealed some current limitations. The key question is whether on- and off-vehicle capabilities can be integrated so that communication, situational awareness, protection, and power-generation requirements can be reduced for the vehicle fleets without significant increases in complexity and cost.

What Congress Can Do

In this monograph, we present a number of strategic, technical, and business practice and process considerations that affect DoD's ability to field combat and tactical wheeled vehicle fleets that meet the country's needs. Some of these take the form of things to pay attention to or do, whereas others frame and, in some cases, constrain DoD's ability to field these vehicle fleets.

One major strategic observation that Congress should consider as it interacts with DoD on the development of vehicle fleets is that predicting future threats over the expected life spans of vehicles now in production is very difficult, and choices must be made and risk accepted due to the impossibility of designing vehicles that are optimal for all future threats. DoD leadership should articulate clearly what rationale it is using in vehicle fleet development (e.g., optimizing vehicles against a specific threat, as in the Cold War, or creating vehicles that are adequate for a spectrum of threats). Given the joint nature of conflict, this rationale should be considered by, if not standard across, each armed service. Congress should consider requiring that DoD present the strategic rationale for these choices *fleet wide*, as well as how each individual proposed vehicle fits within this rationale.

We highlight four classes of technical challenges that currently affect, and for the foreseeable future will continue to affect, the ability of the defense research, development, and acquisition communities to field cutting-edge vehicles that meet the operational requirements of fielded forces: the need for improved protection, power generation, and fuel consumption, and the effect that sensors and networking have on the complexity of modern vehicles, as well as the challenges that come with it. Because these are classes of problems that affect almost every vehicle (and many other systems) that DoD fields, they should be considered as such by Congress. In particular, in its oversight role, Congress should consider taking a role in ensuring that defense programs to address each of these challenges are adequate. This would include, but not be limited to, working with DoD to ensure that these areas are thoroughly addressed. Congress should consider making all four of these areas focal points of its interactions with DoD on research and development, new systems, and modifications to existing systems.

We also identify seven areas in which business practices, processes, and policy changes could significantly enhance the research, development, and acquisition and test and evaluation communities' ability to use resources and time more effectively to accomplish their tasks. Some of these challenges can be addressed—and may be in the process of being addressed or readdressed—by DoD (e.g., how cost estimation is done, how programs are staffed and supported for success, how testing and evaluation are done). Some may require congressional action in the form of guidance, changes to laws, or clarification of congressional intent with a focus on regulations (e.g., adopting ACAT decision practices that more realistically address risk rather than using cost as a proxy for risk).[8] And some, if not all, have cost implications that Congress should factor into the way it oversees vehicle fleet developments (e.g., the rising costs of tactical wheeled vehicles). In all seven cases, Congress may decide that the changes required to make progress will demand that it play some role. Furthermore, in all seven cases, Congress should consider asking for updates and challenging DoD to make or recommend changes.

Finally, a more comprehensive M&S capability and leaders who are empowered to use it well will be essential tools in everything from establishing future requirements to research and development to engineering, program design, and manufacturing. DoD and the services should consider improvements to their already substantial capabilities along the lines presented in this monograph, which will require support and guidance from Congress.

---

[8] John Birkler, Mark V. Arena, Irv Blickstein, Jeffrey A. Drezner, Susan M. Gates, Melinda Huang, Robert Murphy, Charles Nemfakos, and Susan K. Woodward, *From Marginal Adjustments to Meaningful Change: Rethinking Weapon System Acquisition*, Santa Monica, Calif.: RAND Corporation, MG-1020-OSD, 2010.

Acknowledgments

We are indebted to many Army and Marine Corps officials and especially to our sponsor at DDR&E, Howard (Jack) Taylor, who participated actively in facilitating meetings, arranging site visits, and acquiring valuable documentation to support this research. We are also grateful to Spiro G. Lekoudis, director of weapon systems in the Office of the Deputy Under Secretary of Defense for Science and Technology, for his involvement with the research and his participation in the study's interim project reviews. We thank Jillyn Alban, the U.S. Army Tank Automotive Research, Development, and Engineering Center (TARDEC)–Office of the Assistant Secretary of the Army for Acquisition, Logistics, and Technology liaison; Thomas A. Allaire of Headquarters, U.S. Department of the Army (HQDA) G-4; Jeff Bradel from the Office of Naval Research; Roberta Desmond from the Army's Program Executive Office (PEO), Combat Support and Combat Service Support; Tony Desmond, director of systems integration/G-7, PEO Ground Combat Systems; Jeff Gavlinksi from the Army Modeling and Simulation Office, Center for Army Analysis, HQDA G-8; Timothy G. Goddette, director, Combat Sustainment Systems Directorate, Office of the Assistant Secretary of the Army for Acquisition, Logistics, and Technology; David Gorsich, chief scientist at the U.S. Army Research, Development, and Engineering Command (RDECOM)–TARDEC; Keith Hammack, U.S. Navy; Jennifer Hitchcock, acting director of research at TARDEC; Eric Johnson, director of the Wargaming and Simulation Directorate, TRADOC; Reynold (Reed) Skaggs, associate director for plans and programs, Army Research Laboratory; Gary

Trinkle from the Directorate of Plans and Operations and Integration, Requirements Integration Directorate, Army Capabilities Integration Center, TRADOC; Todd Turner, deputy director for air and ground systems, Office of the Deputy Assistant Secretary of the Army, Research and Technology; Don Wyma, deputy director of scientific and technical intelligence in the Office of the Under Secretary of Defense for Intelligence; Mark Young, chief of the Army Modeling and Simulation Office, Center for Army Analysis, HQDA G-8; and Chris Yunker, from the U.S. Marine Corps Office of Combat Development and Integration.

We are also grateful to our reviewers, BG John P. Rose and Elliot Axelband. Their insights and recommendations improved the original manuscript significantly. Our colleague Elliot Axelband went beyond the call of duty in contributing to the revision of Chapter Five.

Abbreviations

ACAT	acquisition category
AoA	analysis of alternatives
APS	active protection system
ATGM	antitank guided missile
C3	command, control, and communication
CDD	capability development document
CPD	capability production document
DDR&E	Director, Defense Research and Engineering
DoD	U.S. Department of Defense
DOTMLPF	doctrine, organization, training, materiel, leadership and education, personnel, and facilities
EFV	expeditionary fighting vehicle
EMD	engineering and manufacturing development
EVMS	earned value management system
FBCF	fully burdened cost of fuel
FCS	Future Combat Systems
FY	fiscal year

GAO	U.S. Government Accountability Office
GCV	ground combat vehicle
HEMTT	Heavy Expanded Mobility Tactical Truck
HMMWV	high-mobility multipurpose wheeled vehicle
HQDA	Headquarters, U.S. Department of the Army
IED	improvised explosive device
IFV	infantry fighting vehicle
ISR	intelligence, surveillance, and reconnaissance
JCIDS	Joint Capabilities Integration Development System
JLTV	joint light tactical vehicle
KPP	key performance parameter
LTAS	Long-Term Armor Strategy
M&S	modeling and simulation
M-ATV	mine-resistant, ambush-protected all-terrain vehicle
MRAP	mine-resistant, ambush-protected (vehicle)
MTVR	medium tactical vehicle replacement
NATO	North Atlantic Treaty Organization
O&S	operating and support
ORD	operational requirements document
PEO	program executive office
PIM	Paladin Integrated Management
PLS	palletized load system

RDECOM	U.S. Army Research, Development, and Engineering Command
RFP	request for proposals
SBCT	Stryker brigade combat team
SDD	system development and demonstration
SEMP	systems engineering management plan
SEP	systems engineering plan
SIL	system integration laboratory
SME	subject-matter expert
SWaP-C	space, weight, power, and cooling
TACOM	U.S. Army Tank-Automotive and Armaments Command
TARDEC	U.S. Army Tank Automotive Research, Development, and Engineering Center
TD	technology development
TRAC	U.S. Army Training and Doctrine Command Analysis Center
TRADOC	U.S. Army Training and Doctrine Command
TRL	technical readiness level
TTPs	tactics, techniques, and procedures
VVA	verify, validate, and accredit

CHAPTER ONE

The Challenges of Providing Appropriate Military Vehicles

The process of research, development, and acquisition that ultimately procures military vehicles and presents them to their ultimate users, soldiers and marines, has historically been difficult. The recent record is consistent with the historical one, although the specific reasons underpinning the challenges of fielding suitable vehicles change with the times. Sometimes, the difficulty lies in translating the threat, such as an enemy antitank guided missile (ATGM), into a design criterion (for example, a protection requirement of so many inches of armor plating). In other instances, the problems have included a mismatch between cost estimates and actual costs (and hence insufficient budgets), creeping or oft-changing requirements, unwarranted faith in immature technologies, or overly ambitious designs. Sometimes, cost growth renders some vehicles unaffordable. Then there are the economics of military acquisitions. Because the military buys relatively small fleets of vehicles of various types, the unit costs and modifications of commercially acquired components tend to be high.

At present, the Army and Marine Corps are in the midst of efforts to develop new vehicles and to recapitalize older vehicle fleets. The Army's new infantry fighting vehicle (IFV), known as the ground combat vehicle (GCV), is in the earliest stages of development and has not yet crossed milestone A in the DoD acquisition process to begin technology development. The joint light tactical vehicle (JLTV) is an Army–Marine Corps and international endeavor; currently, three competitive prototypes are being evaluated. Meanwhile, the Marine Corps' expeditionary fighting vehicle (EFV) has been the subject of journalistic speculation due to a combination of cost growth, schedule slippage, and questions about its operational utility, all of which have raised doubts about its survival after an investment of some $13 billion.

Causes for Concern

Recently, the services have, on occasion, demonstrated difficulties in identifying and responding adequately to appropriate operational requirements. For example, the Army originally sought an 18-ton combat vehicle in the aftermath of 1999's Operation Allied Force, when the Chief of Staff feared that the Army might become "strategically irrelevant" if it could not deploy a suitable rapid-reaction force globally within 96 hours.[1] The Stryker vehicle was procured to satisfy this requirement and to serve as an interim vehicle until the Future Combat Systems (FCS) could be developed. Neither the Stryker nor the FCS family of vehicles, as originally conceived, would have been robust enough to withstand the current threat from improvised explosive devices (IEDs).[2] Indeed, many of the assumptions that lay at the core of those programs do not hold in current conflicts.

Current vehicle designs, most of which date from the Cold War era, had not anticipated such requirements as additional electrical power and protection that resulted from the noncontiguous battlefields

---

[1]   See GEN Eric K. Shinseki, Chief of Staff, U.S. Army, "The Army Transformation: A Historic Opportunity," *2001–02 Army Green Book*, Arlington, Va.: Association of the United States Army, 2001. General Shinseki originally presented the briefing as part of a U.S. Department of Defense (DoD) report to Congress in 1999, when he was the Chief of Staff of the Army. In his statement on the status of forces before the Senate Committee on Armed Services on October 26, 1999, he referred to "soldiers on point for the nation transforming the most respected Army into a strategically responsive force that is dominant across the full spectrum of operations." See also Eric Peltz, John Halliday, and Aimee Bower, *Speed and Power: Toward an Expeditionary Army*, Santa Monica, Calif.: RAND Corporation, MR-1755-A, 2003.

[2]   "Gates expressed a specific concern that the portion of the FCS program to field new manned combat vehicles did not adequately reflect the lessons of counterinsurgency and close quarters combat in Iraq and Afghanistan" (Office of the Assistant Secretary of Defense for Public Affairs, "Future Combat System [FCS] Program Transitions to Army Brigade Combat Team Modernization," news release, No. 451-09, June 23, 2009).

to which Army and Marine Corps units are currently deployed. The nonlinear, irregular distribution of brigade and battalion formations means that there is no longer a relatively more secure rear area, an expectation that the enemy will be in front of advancing U.S. forces, or an assumption that all units—and therefore all vehicles—will face similar threats. Thus, simple logistics vehicles that once differed little from their commercial counterparts now require enhanced situational awareness (i.e., a sense of one's surroundings and the location of the enemy and other friendly forces in an area) and protection that are similar in important ways to those of the ground combat vehicles that purposely engage the enemy. The imperative for situational awareness requires these vehicles to have electrical power to support radios, sensors, and computer-based systems that deliver the necessary awareness—performance criteria not anticipated when the vehicles were designed. And the need for protection from IEDs demands that this class of vehicles have armored cabs, at a minimum.

A number of studies, many of which are listed in the bibliography of this monograph, and the judgment of experts at RAND and in the broader community of defense analysts have raised concerns about tactical wheeled and ground combat vehicle programs. Some of the specific concerns include "requirements turbulence" (the frequent changing of requirements), inadequately defined requirements, a lack of adequate systems engineering and systems integration in the vehicle's development, and reliance on immature technologies in the vehicle's design (which exposes the program to risk if the technology in question matures more slowly than anticipated). These studies and experts also point to business risks, such as those arising from overly optimistic estimates of a program's schedule and likely costs. These arguably controllable business risks are exacerbated, and in some cases overshadowed, by funding instability.[3] This, in turn, undermines a program's ability

---

[3] There are several potential sources of funding instability. Generally, funding instability results from service decisions to distribute available funds across a series of programs such that some programs receive robust funding and others are maintained on relatively meager budgets. In subsequent years, the services adjust funding to reflect their expectations about program performance, a practice that causes the underfunded programs to hedge against further reductions.

to maintain its schedule and results in additional costs, among other things.

Signs of Progress

Although the recent record raises cause for concern, there are reasons to expect improvements in the services' vehicle programs.

In an attempt to sharpen the requirements process, U.S. Army Training and Doctrine Command (TRADOC) created Task Force 120 (since closed down) to refine future force concepts. The goal of this effort was to help more closely align vehicle performance criteria with the demands faced in terms of the enemy, weather, and terrain.

The Army Vice Chief of Staff's Capability Portfolio Reviews were created at the direction of the Secretary of the Army to conduct an Army-wide, all-components revalidation of the operational value of Army requirements within and across capability portfolios to existing joint and Army warfighting concepts. The intent of this revalidation is to eliminate redundancies and to ensure that funds are properly programmed, budgeted, and executed against the programs that yield the most value to the Army.[4]

The Army also instituted configuration steering boards for major acquisition programs and limits on configuration changes that could lead to substantial cost growth or production slips.[5] Similarly, the Marine Corps created gate reviews in its major acquisition programs "to improve governance and insight, ensure alignment between capability requirements and acquisition, improve senior leadership decision making, and gain better understanding of risks and costs."[6]

---

[4]   Army Force Management School, "Force Management Update," *AFMS News Letter*, July 2010.

[5]   "Highlights of the FY2009 NDAA," *Federal Contracts Report*, Vol. 89, No. 22, June 10, 2008.

[6]   Defense Acquisition University, "Naval Gate Review Process," briefing, March 19, 2009.

The Army has also taken steps to reduce the risk associated with technology. Program managers indicated[7] that they are less likely to consider technologies that have not achieved a technology readiness level (TRL) of 6 or better during the technology development phase (that is, between milestones A and B) of the programs.[8] As a result, technology innovation, if it is to occur, must take place either early in the development or outside of programs of record. In addition, the service and its contractors employ various techniques, such as systems engineering management plans (SEMPs) and systems engineering plans (SEPs), to overcome reported shortcomings in program systems engineering. Recent vehicle designs often feature open architectures, which, in theory, will make it easier for the vehicles to accommodate subsequent improvements by reserving space, weight, and power to support new capabilities. Finally, the Army has made a deliberate effort to benefit from technology advances by inserting upgraded technology into the current fleets of vehicles during the reset process as these vehicles are reconditioned upon their return from duty overseas.[9]

The services have also sought to manage business risk. The Army, for example, has committed to knowledge-based acquisition,[10] which predicates movement along the program's schedule on having acquired the requisite knowledge to manage the risks inherent in the step forward. Thus, for example, a program manager would not move to low-rate initial production until he or she was confident that the decision was warranted based on testing of the prototypes. The JLTV program

---

[7] In comments made during the project's workshop, July 29, 2010.

[8] Paul Rogers, Executive Director of Research, U.S. Army Tank Automotive Research, Development, and Engineering Center, U.S. Army Research, Development, and Engineering Command, "FCS Technology Insertion and Transition," briefing presented at the eighth annual Science and Engineering Technology Conference, North Charleston, S.C., April 18, 2007.

[9] Rogers, 2007.

[10] Michael J. Sullivan, Director of Acquisition and Sourcing Management, U.S. Government Accountability Office, *Defense Acquisitions: Opportunities and Challenges for Army Ground Force Modernization Efforts*, testimony before the Senate Committee on Armed Services, Subcommittee on Airland, Washington, D.C.: U.S. Government Accountability Office, GAO-10-603T, April 15, 2010.

uses "knowledge point reviews" to determine whether requirements are achievable in the engineering and manufacturing development (EMD) phase of the program.[11] As part of the 2005 Base Realignment and Closure recommendation, the Technical Joint Cross Service Group recommended that the Army and the Marine Corps consolidate facilities into a joint ground vehicle center for development and acquisition.[12] Recently, in part as a response to the 2005 recommendation, the services have begun to put into place a formal construct, referred to as the Joint Center for Ground Vehicles, to help institutionalize system integration, collaboration, and portfolio management for the ground vehicle community. The center includes parts of the U.S. Army Tank-Automotive and Armaments Command (TACOM) Life Cycle Management Command; Program Executive Office (PEO) for Ground Combat Systems; PEO Combat Support and Combat Service Support; PEO Integration; the U.S. Army Tank Automotive Research, Development, and Engineering Center (TARDEC); and the U.S. Marine Corps PEO Land Systems and Marine Corps Systems Command.[13]

Program managers also remarked favorably on the potential for "incremental acquisition," whereby they field a vehicle in successive generations, to occur more often than in the past and with fewer vehicles, with each generation reflecting improvements over its predecessor generations. This approach provides a means to manage costs, react to changing requirements, and facilitate technology insertion within the strictures imposed by the Joint Capabilities Integration Development System (JCIDS) process.[14] Finally, the services pointed to rapid acqui-

---

[11] COL John S. Myers, Project Manager, Joint Combat Support Systems, "Joint Combat Support Systems," briefing to the National Defense Industrial Association Tactical Wheeled Vehicles Conference, Monterey, Calif., February 9, 2010.

[12] Technical Joint Cross Service Group, *Analysis and Recommendations*, Vol. 12, May 19, 2005.

[13] Mike Viggato, "Army and Marines Establish the Joint Center for Ground Vehicles," *Accelerate*, Summer 2010.

[14] Comments made during the project's workshop, July 29, 2010.

sition and fielding initiatives as efficient and reliable ways to produce capabilities for the deployed forces.[15]

We discuss these and related issues in later chapters.

Congressional Concerns

In Section 222 of the National Defense Authorization Act for Fiscal Year 2010 (Pub. L. 111-84), Congress directed the Secretary of Defense to hire an independent agency to develop a report that would help Congress fulfill its role in overseeing how the armed services produce vehicles suitable for anticipated near- and mid-term threats. Specifically, Congress asked for an independent assessment of current, anticipated, and potential research, development, test, and evaluation activities related to the modernization of DoD's ground combat vehicle and armored tactical wheeled vehicle fleets and to submit interim and final reports on the assessment to the congressional defense committees:

> (A) A detailed discussion of the requirements and capability needs identified or proposed for current and prospective combat vehicles and armored tactical wheeled vehicles.

> (B) An identification of capability gaps for combat vehicles and armored tactical wheeled vehicles based on lessons learned from recent conflicts and an assessment of emerging threats.

> (C) An identification of the critical technology elements or integration risks associated with particular categories of combat vehicles and armored tactical wheeled vehicles, and with particular missions of such vehicles.

> (D) Recommendations with respect to actions that could be taken to develop and deploy, during the ten-year period beginning on the date of the submittal of the report, critical technology

---

[15] The most notable example for vehicles is the mine-resistant, ambush-protected (MRAP) vehicle program. See U.S. Government Accountability Office, *Defense Acquisitions: Issues to Be Considered as DoD Modernizes Its Fleet of Tactical Wheeled Vehicles*, Washington, D.C., GAO-11-83, November 5, 2010e, for a discussion of how this helped enable the successful fielding of the MRAP all-terrain vehicle (M-ATV).

capabilities to address the capability gaps identified pursuant to subparagraph (B), including an identification of high-priority science and technology, research and engineering, and prototyping opportunities.[16]

Research Methodology

This study was funded at a level that limited its research design. We had to modify our proposed approach to consider a smaller sample of vehicle types. We could not examine requirements and capability needs as Congress anticipated we would in paragraph (A) above, although we address them in a limited fashion in Chapter Three. We could not examine all potentially valuable technology elements anticipated in paragraph (C) and instead had to focus on broader technology areas (e.g., power generation, protection). Nor could we conduct modeling and simulation (M&S) in support of the research effort, although we have devoted Chapter Five to a discussion of how M&S could be improved in support of vehicle development.

The research approach focused on a selection of combat and tactical wheeled vehicles that served as proxies for different classes of vehicles (e.g., heavy truck, main battle tank) and that were at different stages of development. (The specific vehicles are discussed in the next section.) This structure provided us with an opportunity to examine a full range of issues as represented by a relatively small sample of vehicles.[17] Ultimately, we adopted an approach with four principal lines of inquiry: (1) scrutinize individual example vehicles; (2) conduct a literature review; (3) observe military sites, facilities, and activities associated with the research, development, and procurement of the vehicles in question; and (4) host a workshop to collect the views of service officials and subject-matter experts (SMEs). Ultimately, we synthesized

---

[16] Public Law 111-84, National Defense Authorization Act for Fiscal Year 2010, October 28, 2009, Sec. 222.

[17] Resource constraints prevented us from examining this problem in a more comprehensive context.

what we learned from each line of inquiry to reach insights bearing on the congressional questions.

Example Vehicles

The example vehicles were selected with the assistance of our sponsors in the Office of the Director, Defense Research and Engineering (DDR&E). Vehicles were organized into a first tier, which would receive the greatest attention, and a second tier, subject to less scrutiny. Tier 1 included the Army GCV; the JLTV sponsored by the Army, Marine Corps, other services, and foreign partners; the Marine Corps EFV and medium tactical vehicle replacement (MTVR); and the Army's Heavy Expanded Mobility Tactical Truck (HEMTT). As noted, the GCV, JLTV, and EFV are currently in development. The MTVR and HEMTT are currently in the force and have established histories of performance. Tier 2 vehicles are all currently serving in the force and are undergoing modernization or recapitalization. They include the M1 Abrams tank, the Stryker family of eight-wheeled armored cars, and the M109A6 Paladin 155 self-propelled howitzer and its support vehicle (the Paladin Integrated Management, or PIM, initiative, which will move the gun system from the early 1960s–era M109 chassis to the M2 Bradley chassis).

Considered together, the two tiers provide examples of wheeled vehicles, tracked vehicles, tactical vehicles, and combat vehicles, as well as fielded vehicles and vehicles that are still in development, as shown in Table 1.1.

Literature Review

The literature review emphasized the requirements documents for the vehicles in question. In addition to analyzing the requirements documents, we also collected reports on the vehicles and the research, development, and acquisition processes written by government regulatory and watchdog agencies, including the U.S. Government Accountability Office (GAO), the Congressional Budget Office, and the Congressional Research Service. We also reviewed relevant studies conducted by federally funded research and development centers, including those operated by RAND and the Institute for Defense Analyses. We

Table 1.1
Taxonomy of Example Vehicles

Vehicle	Wheeled	Tracked	Tactical	Combat	Fielded	In Development
GCV	?	?		x		x
JLTV	x		x			x
EFV		x		x		x
MTVR	x		x		x	
HEMTT	x		x		x	
Abrams		x		x	x	
Stryker	x			x	x	
PIM		x		x		x

NOTE: The information on the GCV is based, in part, on the initial request for proposals (RFP) for the GCV, which has been withdrawn. This information may change when the new RFP is issued.

reviewed program management documents and briefings, materials from the Defense Acquisition University, and documents provided by the Army and Navy science and technology communities. We also consulted the publicly available reports and other documentation from the Army Research Laboratory, the Joint IED Defeat Organization, and the Marine Corps PEO Land Systems and Capabilities Development Directorate. Finally, we read unit after-action reviews for indications of vehicle performance and shortcomings.

Site Visits

Members of the research team visited TARDEC/TACOM, the Army Research Laboratory, the Joint IED Defeat Organization, and the Marine Corps' PEO Land Systems and Capabilities Development Directorate at Quantico, Virginia. During these visits, team members received briefings from program executive officers, program managers, and their staffs; met with officials engaged in research, development, testing, and evaluation; interviewed operational test and evaluation personnel; and spoke with members of the combat development communities in both services. We also interacted with members of the Intelligence Community and received briefings on the current threat to the tactical wheeled and ground combat vehicle fleets. We visited the Office of Naval Research and also met with members of the Navy science and technology community.

The Project Workshop

RAND hosted a project workshop in its offices in Arlington, Virginia, on July 29, 2010. Attendees included SMEs on how the services develop requirements and the vehicles that populate the study. Program managers from both the tier 1 and tier 2 programs attended, as did members of the combat developer, science and technology, and operational test and evaluation communities and other members of the service acquisition communities. The Office of the Under Secretary of Defense for Intelligence provided threat information for the workshop. The science and technology community also supported the workshop, sending attendees to the technology breakout group and helping the research team understand the state of the art in key technology domains. Finally, the M&S community's participation enhanced our understanding of its current capabilities and the future potential to do more in support of vehicle research, development, and acquisition.

Organization of This Monograph

The monograph contains six chapters. Chapter Two discusses how operational requirements for forces with specific capabilities are translated into program requirements for vehicles that provide the capabilities that, in turn and collectively, define what the U.S. military combat and tactical wheeled vehicle fleets will look like in the future. The chapter provides a brief conceptual overview that emphasizes the challenges of doing this well, introduces the concepts of *operational* and *program* requirements that vehicle programs must meet to be successful, and foreshadows the discussion in Chapter Five on the need for a suite of well-synchronized analytical tools.

Chapter Three summarizes requirements for the vehicles examined in the study. The substance of Chapter Three is a qualitative analysis to describe how well the requirements for the vehicles in question prepared them for anticipated operating environments, where those criteria left the vehicles vulnerable, and where capability gaps exist as a result.

Chapter Four illustrates the interdependencies among vehicle requirements: how overall trafficability and turning radius affect vehicle size and weight, how protection requirements can affect weight, and how weight, in turn, can impose demands on suspension and propulsion systems. The description in Chapter Four illustrates how new requirements in one dimension of vehicle design can affect the others, sometimes negatively, and how accommodating some design points can require difficult trade-offs. Finally, the chapter describes how trade-offs occur in the context of a specific set of anticipated circumstances and how, when reality diverges from the anticipated circumstances, vehicle performance can suffer as a result.

Chapter Five addresses M&S from a perspective that differs significantly from that of the other chapters. Because we were unable to undertake the M&S that might have revealed how simulation-based trade-offs could bring new precision to vehicle designs, Chapter Five proposes how M&S might be deployed to improve vehicle designs. In the course of this description, Chapter Five also incorporates other insights from site visits, the literature review, and the study's workshop that, if adopted, could increase the contribution of M&S to vehicle research, development, and acquisition.

Chapter Six presents our observations and conclusions. The chapter synthesizes some of the observations from the various lines of inquiry that animated the study. It also employs deduction and inference to reach some of the observations. These observations are building blocks of information that, when properly assembled in Chapter Six, inform the study's conclusions and address Congress' concerns specifically.

CHAPTER TWO

Operational and Program Requirements

One way to summarize the questions posed by Congress in Section 222 of the National Defense Authorization Act for Fiscal Year 2010 (Pub. L. 111-84) is, "Will DoD's planned vehicle fleet meet the nation's requirements over the coming decade, and if not, what can Congress do about it?"[1] To answer that question, one needs to answer several important subsidiary questions, including the following:

- What capabilities are needed on the battlefield?
- How are these capabilities translated into requirements for combat and tactical wheeled vehicles?
- What are the challenges with fulfilling these requirements?
- What can Congress do about it?

This chapter examines a set of considerations that are important for answering the first two of these questions, and subsequent chapters look at the other two.

Requirements

The word *requirement* is used in a few different ways with respect to DoD capabilities and systems in general. According to Joint Publication 1-02, a military requirement is an "established need justifying

---

[1] This restatement was briefed to appropriate congressional defense committee staff members in October 2010 and was met with approval.

the timely allocation of resources to achieve a capability to accomplish approved military objectives, missions, or tasks."[2] When DoD refers to "requirements" in the context of producing vehicles for its military forces, it is typically referring to performance criteria for the equipment that it is in the process of developing and procuring. The performance criteria tend to be shaped by DoD's assumptions about its adversaries and the wars its forces will fight. These requirements are time-bound; they reflect the considerations influencing DoD thinking at the time, including assumptions about the future.

In this sense, requirements are levied on the services (and other force providers) to provide capabilities to warfighters.[3] These requirements for capabilities can be short-term, such as in response to urgent operational needs identified during current operations or in anticipation of impending operations, or longer-term, as indicated by joint force or service concept development processes. While these requirements stemming from urgent needs or concepts for future wars will have different implications for the acquisition community, conceptually, they are similar. We call them *operational requirements*.[4] While the focus of this monograph is on vehicles, operational requirements are usually levied in terms of capabilities that unit types should have.

Once operational requirements for capabilities are defined, the armed services are tasked to provide them. In theory, they can do this by making adjustments to one or more of the major institutional systems for which they are responsible. These systems are summarized in

---

[2]   U.S. Joint Chiefs of Staff, *Department of Defense Dictionary of Military and Associated Terms*, Joint Publication 1-02, Washington, D.C., April 12, 2001, as amended through September 30, 2010, p. 296.

[3]   U.S. Special Operations Command is an example of a nonservice force provider. We focus almost entirely on the Army and Marine Corps in this monograph, so further references will be to the armed services only, with the understanding that there are other force providers.

[4]   We did not attempt to independently identify operational requirements or determine whether current operational or programmatic requirements (defined later in this chapter) were adequate. Rather, we accepted DoD's statements of requirements as given.

the acronym "DOTMLPF."[5] In general, force development procedures favor nonmaterial changes to fulfill new requirements for capabilities, as they tend to be cheaper and can be made more quickly. However, once the decision has been made to meet an operational requirement with a material solution (we call such a material solution a "system"), the combat development community is charged with translating operational requirements related to desired capabilities for a *unit* into *program requirements* that define the characteristics that a *system* (in our case, combat or tactical wheeled vehicles) should have.[6]

Program requirements must capture performance criteria for a host of vehicle characteristics—for example, their mobility; ability to provide power to radios, sensors, and other embedded technologies; level of protection against enemy weapons; and weapon performance. Performance (recently emphasizing power), protection, and payload have become the "iron triangle" of vehicle requirements. These requirements interact with each other, often in complex ways. Protection, for example, can add weight, which, in turn, undermines mobility. Power generation, insofar as it also produces heat and noise, can compromise protection by creating heat and noise signatures that an enemy could detect. However, insofar as it activates and operates an active self-defense system, power generation improves protection.

In reality, combat developers work to define these requirements cooperatively with the acquisition community, with the former acting as the proponent for what is desired and the latter acting as the proponent for what is possible and affordable. Those in both communities

---

[5] DOTMLPF stands for doctrine, organization, training, materiel, leadership and education, personnel, and facilities.

[6] In the Army case, TRADOC provides a community of combat developers tasked with articulating requirements for new vehicles. In the Marine Corps, the PEO Land Systems and Capabilities Development Directorate at Quantico performs the same basic function. The members of the combat developer communities are generally not scientists or engineers but experienced officers with a profound understanding of recent and ongoing combat operations and how future concepts are developed (i.e., operational requirements) and their implications for vehicle performance (i.e., program requirements). This process of translating unit operational requirements into system program requirements is a complex task, as systems do not operate as stand-alone entities on the battlefield. We return to the subject of how to improve this translation process when we discuss M&S in Chapter Five.

interviewed for this research indicated that program requirements are best defined by an active collaboration between the two. This process is depicted in Figure 2.1. In the figure, operational requirements are shown in the topmost rectangle, and the vertical arrows indicate that they change over time (indeed, currently, they change relatively rapidly as the result of the changing demands of ongoing conflicts). The lower left side of the figure shows that operational requirements are translated into program requirements, which, in turn, are taken up by the research, development, and acquisition communities. These communities are charged with developing and building the specified systems. Note that program requirements also have vertical arrows, which indicate that they are not set in stone and may change during the development of a system (e.g., current changes in the GCV concept) and almost certainly will over the life span of a system. Then, the system as delivered (on the right side of the figure) may meet, exceed, or fail to meet the specified system capabilities. Because of the long lead times in developing major systems, the system may or may not meet operational

Figure 2.1
Requirements and System Risk

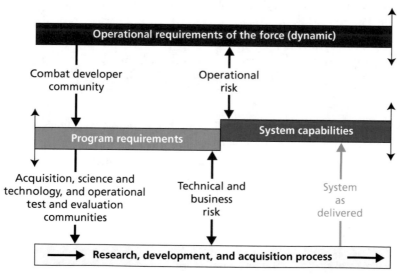

requirements. Finally, the extent to which it fails to meet operational requirements can be characterized as "risk," of which we highlight two types: operational and program (which is further decomposed into technical and business risk in the figure).

It is important to note that operational requirements exist only with respect to the capabilities needed in a current or anticipated conflict. We discuss capabilities in more detail in the next chapter. We note that several types of conflict are recognized by DoD,[7] each of which generates operational requirements that may or may not be compatible with the operational requirements of other types of conflict. A brief overview of Army modernization trends from the mid-1990s to the mid-2000s illustrates this well.

Prior to the conflict in Kosovo, the Army's principal ground combat vehicle fleet consisted of very heavy main battle tanks (the M1 Abrams class), infantry fighting vehicles (the M2 Bradley class), and assorted other heavy weapon systems. They require significant amounts of strategic lift to move and large logistical efforts to support, and they were optimized to fight peer competitors, for which they would rely on heavy frontal armor, sophisticated and accurate direct and indirect fire systems, and tactical mobility to succeed. Following the Army's struggles to deploy a single, modest task force (Task Force Hawk) in support of Operation Allied Force in Kosovo in 1999, then–Chief of Staff of the Army GEN Eric Shinseki set the Army on a modernization path that emphasized strategic mobility. The backbone of this effort was the perceived need for a C-130–deployable family of ground combat vehicles that would become known as the "Future Combat Systems," or FCS. These vehicles were to rely on almost-omniscient intelligence, fully networked sensors and shooters, and great tactical mobility rather than heavy frontal armor to defend themselves against and defeat enemies ("see first, understand first, act first, and finish decisively").[8] Less

---

[7] See, for example, Headquarters, U.S. Department of the Army, *Operations,* Field Manual 3-0, Washington, D.C., February 27, 2008, Chapter Two (on the spectrum of conflict and operational themes).

[8] BG David Fastabend, Deputy Director, U.S. Army Training and Doctrine Command Futures Center, "The Army in Joint Operations: The Army's Future Force Capstone Concept 2015–2024," video, July 25, 2005.

than five years later, the Army and Marine Corps found themselves embroiled in wars in Iraq and Afghanistan that would see a demand (requirement) for a very different type of vehicle than the envisioned 18-ton FCS or the Cold War legacy vehicle fleet. Here, threats from IEDs and fighters who hid among the population were the principal threats, and protection for all vehicles, including tactical wheeled vehicles (heretofore not well protected or networked), particularly from underbelly explosives, became the order of the day.

We note that over the span of less than a decade, these three concepts produced vehicle requirements that were incompatible in important ways. For example, optimizing vehicles for a conflict in which a near-peer competitor will primarily use direct-fire weapons in large numbers argues for combat vehicles with low silhouettes, heavy frontal armor, fast acceleration, cross-country mobility, and the ability to engage in direct and indirect fire on a large scale. They are large, heavy vehicles with good tactical mobility that require significant amounts of strategic lift and long lead times to deploy, and they have large logistical footprints. In contrast, a force that can deploy capable units of action in medium-lift aircraft will necessarily be light and have smaller logistical requirements. They will be unable to sustain the same direct-fire engagements as a heavier force, however. Finally, vehicles that will be effective in protecting soldiers from underbelly IEDs (our current emphasis) need high silhouettes (to provide additional distance between the blast and the vehicle) and V-shaped hulls that add height and weight to the vehicle, and they do not require thick frontal armor. In particular, vehicles cannot be designed to provide low silhouettes and high silhouettes, significant underbelly and frontal armor and good tactical mobility (due to weight limits), and great on-vehicle protection while remaining lightweight and strategically mobile; incorporating all of these characteristics collectively is a physical impossibility.[9]

---

[9]   While this example emphasizes ground combat vehicles, a similar example that compares the requirements for tactical wheeled vehicles with expectations for the threat geography of the battlefield would point out similar lessons. In particular, if battlefields have relatively secure rear areas, then tactical wheeled vehicles do not need the enhanced protection of MRAPs or advanced networking equipment that provides real-time situational awareness.

This so-called "iron triangle" of trade-offs among performance, protection, and payload is discussed at greater length in Chapter Three.

This example illustrates one particularly important point for this research: For developers to translate operational requirements into program requirements and then field a fleet of vehicles that will meet those operational requirements, strategists must provide the combat developers with an adequate concept for future operations a decade or more out, unless the United States intends to buy more than one fleet of vehicles for its armed forces. Recent history indicates that this is a real challenge, and current budget concerns indicate that purchasing, training on, modernizing, and maintaining multiple vehicle fleets is not affordable. To the extent possible, choices must be made and risk accepted, with the understanding that rapidly evolving requirements—or requirements for entirely new classes of vehicles (e.g., the MRAP)—require very responsive combat developer and acquisition systems, as well as generous resources. Alternatively put, such changes put tremendous stress on acquisition systems and introduce both operational and technical/business (i.e., program) risk.

Successful Systems

Now that we have defined the operational and program requirements and risk, we can discuss more carefully how they affect whether a system is "successful." Meeting both operational and program requirements is a "necessary" element for success in the sense that if a system fails to meet requirements in either category, it can be categorized as a failure.

The first and most critical determinant of success is whether a system meets operational requirements. This, in turn, is reflected in the characteristics it must have to actually be effective. If the combat developer community has adequately defined program requirements, and if they have been met, then the system is successful as delivered. If not,

---

Both are requirements on a battlefield in which enemies exist throughout and there are only small islands of real security.

operational success may still be possible to some degree (for example, if it is an improvement over existing systems or provides needed capabilities beyond those currently in the force).

A second and important criterion—or set of three distinct criteria—is whether a system meets (1) program requirements (2) on time and (3) within budget. Note that a system can be a programmatic success and an operational failure if it satisfies these three criteria but does not meet the (possibly changing) requirements of the force (though this would mean that program requirements do not adequately reflect operational requirements). For example, it could meet stipulated weight requirements but be too heavy for roadways and bridges where it is deployed. Note that we can break programmatic success (and risk) down into technical success and business success (and risk):

- *Technical success* occurs when a program delivers a system that meets program requirements. The core element of technical success is adequately addressing the technical and engineering challenges of manufacturing a system that meets program requirements. *Technical risk* is the risk incurred in accomplishing these tasks.
- *Business success* can be defined as occurring when a program delivers a system on time and within budget. Business success has many parts, among the most important of which are properly designing a program's efforts (including schedule, vendor selection, and coordination of all major efforts), securing stable funding (including plans based on accurate cost estimates), and ensuring that the human resources are in place to properly manage and execute the program. Many important players are responsible for business success, including program executive officers and program managers responsible for the "tactical" aspects of setting up, overseeing, and executing a program. A second group includes department and service senior acquisition officials, who set acquisition priorities, provide higher-level oversight, and secure funding and (in cooperation with personnel managers) human resources. *Business risk* is the risk incurred in accomplishing these tasks.

In addition to these necessary elements of success, there are constraints on programs that may cause desired systems to be unfeasible. Such constraints include business processes (e.g., the DoD 5000-series regulations), statutes, and the state of technology. For example, a program may not be feasible because business processes do not permit actions that would allow it to be completed on time or within budget, or technologies may not mature as quickly as anticipated to deliver the necessary capabilities within the planned time horizons. In both cases, it is the responsibility of the acquisition and combat development communities to let service leaders know whether desired programs are feasible. In particular, given a set of requirements, program managers are responsible for designing programs within these known constraints and letting more senior officials know if they are not executable.

Note, too, that most of the necessary elements have to do with the relationship of some task performed by a major player and some requirement. For example, do the requirements as stipulated for a system by combat developers meet the real requirements (missions) of the force? If not, the system will fail for operational reasons. Do the technical and engineering aspects of the system meet the stipulated requirements for the program? If not, the system will fail for technical reasons. Will the program deliver a system within stipulated time frames to meet the requirements of the force? If not, the system will fail for business reasons.

Efforts to Mitigate Risk

The Army and Marine Corps have adopted a number of approaches to mitigating risk. This section summarizes those that are most salient to the fleets of vehicles considered in this study.

Requirements Refinement

Requirements "creep" and "turbulence," according to the project's workshop participants, have been major sources of risk for both the Army and Marine Corps. As a result, both services have taken steps to sharpen their requirements processes and reduce the likelihood of creep and turbulence.

The Army began the Vice Chief of Staff Portfolio Review Process in 2010. Two portfolios are of particular interest for our purposes: the tactical wheeled vehicle portfolio and the ground combat vehicle modernization portfolio. They provide comprehensive reviews of their subject areas, including requirements and investment strategies, and involve key participants from the Army staff, PEOs, project managers, the Army Capabilities Integration Center, and the broader research, development, and acquisition communities.[10]

Configuration steering boards represent another tool for ensuring requirements stability for the military departments' acquisition programs. As originally conceived by Congress,[11] such configuration steering boards would be

> responsible for reviewing any proposed changes to program requirements or system configuration that could have the potential to adversely impact program cost or schedule and for recommending changes that have the potential to improve program cost or schedule in a manner consistent with program objectives.[12]

The Marine Corps also makes use of gate reviews. The Marine Corps *Operational Test and Evaluation Manual* describes the expected effects:

> The Gate Review process helps ensure alignment between capability requirements and acquisition while improving senior leadership visibility into program risks and costs throughout the development cycle. [The Department of the Navy] has adopted the Probability of Program Success (PoPS) approach, used in con-

---

[10] Jim Rowan, Deputy Commandant, U.S. Army Engineer School, "Update to Emeritus Leaders," briefing, September 29, 2010.

[11] U.S. Senate, Committee on Armed Services, *National Defense Authorization Act for Fiscal Year 2009*, Report No. 110-335, Washington, D.C.: U.S. Government Printing Office, May 12, 2008, sec. 803, 2009.

[12] U.S. Senate, Committee on Armed Services, 2009, sec. 803.

junction with Gate Reviews, to assess and monitor the health of naval [including Marine Corps] acquisition programs.[13]

Figure 2.2 illustrates when these reviews occur in the Marine Corps acquisition process. As the figure suggests, gate reviews occur twice during the technology development phase (yellow circles 4 and 5 in the figure) and twice during engineering and manufacturing devel-

Figure 2.2
Marine Corps Gate Reviews in the Acquisition Process

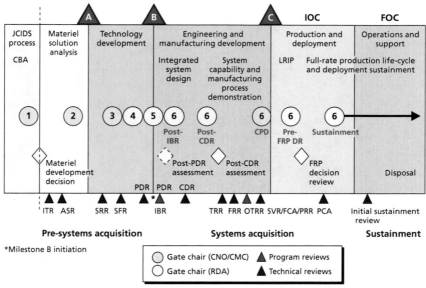

SOURCE: U.S. Marine Corps Operational Test and Evaluation Activity, 2009, p. 1-5.
NOTE: ASR = acquisition strategy report. CDR = critical design review. CMC = Commandant of the Marine Corps. CNO = Chief of Naval Operations. DR = development review. FCA = functional configuration audit. FRP = full-rate production. FRR = functional readiness review. IBR = integrated baseline review. ITR = initial technical review. LRIP = low-rate initial production. OTRR = operational test readiness review. PCA = physical configuration audit. PRR = production readiness review. RDA = research, development, and acquisition. SFR = system functional review. SRR = system readiness review. SVR = system verification review. TRR = technical readiness review.
RAND MG1093-2.2

---

[13] U.S. Marine Corps Operational Test and Evaluation Activity, *Operational Test and Evaluation Manual*, v1.1, October 1, 2009, p. 1-4.

opment (yellow circles labeled 6). Each of the gates is qualitatively somewhat different from the others, but in general, they all seek to establish clear requirements for test and evaluation of the vehicle under development.

Technology Risk Mitigation Efforts

One of the positive developments in the Army acquisition community from the FCS program is a preference for more mature technologies.[14] In other words, program managers prefer to embrace relatively mature technologies, typically those already at TRL 6 or better, as a program enters the technology development phase. This preference for more mature technologies increases the chance that all technologies will have evolved significantly by the end of the technology development phase and be mature enough for low-risk engineering and manufacturing development.

SEMPs have become more detailed and commonplace over the years and help program managers monitor their contractors. As the Defense Acquisition University website describes them,

> The SEMP shall describe the contractor's system engineering process as it is proposed to be applied to the definition of system design and test requirements during the contractual effort. It shall include the system engineering required to define system performance parameters and preferred system performance parameters and preferred system configuration to satisfy the contractual requirements; the planning and controls of the technical program tasks; and management of a totally integrated effort of design engineering, test engineering, logistics engineering and production engineering to meet cost, technical performance, and schedule objectives. The SEMP will be used to understand and [evaluate] the contractor's engineering work efforts as part of the contract monitoring process.[15]

---

[14] A consensus that emerged during the project's workshop, June 29–30, 2010.

[15] Defense Acquisition University, "Data Item Description (DID) System Engineering Management Plan (SEMP)," web page, October 11, 2002.

The SEMP thus affords program managers detailed insights into the systems engineering processes that their contractors propose to employ and in the process helps manage technical risk. SEPs are more recent, internal program tools that complement the SEMP in reducing technical risk. DoD policy requires that SEPs be approved by the milestone decision authority in conjunction with each milestone review and integrated with program acquisition strategy.[16] Program managers update their SEPs as their programs evolve to offer

> a common reference to achieve stakeholder insight regarding a program's planned technical approach. It documents technical managers' understanding of how the program will accommodate and balance cost, schedule, performance, and sustainment requirements and constraints; the expected products of systems engineering activities; and how these products will contribute to program decisionmaking.[17]

Open-architecture designs like those favored for the GCV may represent another technical risk management practice influenced by the FCS experience. The virtues of open-architecture designs for limiting technical risk are threefold. First, vetronics, the ground vehicle equivalent of avionics, provides a data bus and data bus standards that allow rapid substitution of different electronic components as may be required during vehicle upgrades, lowering the cost of the upgrades in the process. Second, open-architecture designs provide surplus size, weight, power, and cooling (SWaP-C) capabilities aboard the vehicle. Thus, if technologies are slow to evolve and require additional SWaP-C to make them tenable aboard the vehicle, the design can accommodate the technology's needs. Second, open-architecture designs go hand in hand with the "buy fewer more often" strategy for buying vehicles. According to this school of thought, especially for vehicles that represent major design departures from those they replace (e.g., JLTVs over HMMWVs), prudence dictates buying a relatively small initial lot,

---

[16] Defense Acquisition University, "Systems Engineering Plan (SEP)," *ACQuipedia*, April 19, 2005.

[17] Defense Acquisition University, 2005.

then modifying the second lot in response to feedback from the field (incremental acquisition, discussed later).[18] This purchasing approach fits neatly with open-architecture designs because such designs have a relatively more abundant SWaP-C with which to accommodate the feedback from the field. Taken together, open-architecture designs and "buying fewer more often" reduce the technical risk associated with a class of vehicles.

Another practice that reduces technical risk involves the periodic use of technology insertion packages. These packages provide upgrades to a unit's equipment, if available, as the unit undergoes the reset phase of the Army Force Generation process. Although there is some risk in any technology insertion process, if done correctly, mission risk is reduced because the reset equipment is more mission-capable than before reset. Technology insertion "ensures that mature technological solutions increase readiness, reduce life cycle costs, and reduce the logistics footprint. New capabilities must be crafted to deliver technologically sound, sustainable, and affordable increments of militarily useful capability."[19]

Competitive prototyping is also associated with minimizing technical, cost, and schedule risk. The JLTV program offers a current example, in which three teams of contractors have submitted prototypes. The next chapter treats the JLTV in more detail, but the point for the purposes of technical risk management is that competing prototypes may reveal different approaches to delivering the performance capabilities that the services seek. Novel approaches may avoid the technical risk inherent in other designs that seek leverage through less mature technologies. At a minimum, multiple prototypes are likely to present the art of the possible as a spectrum, something single prototypes alone cannot do. In addition, prototypes lower costs and schedule risk because they provide a firmer basis for estimation of future cost and schedule and, in this way, lower program risk. Programs with competitive prototypes enable all competitors to better estimate future cost

---

[18] This practice may also have negative effects, such as increasing engineering workload and unit costs.

[19] Defense Acquisition University, "Technology Insertion," web page, May 20, 2003.

and schedule and therefore give DoD a better basis for its selection of a prototype to take to the next step in the acquisition process.

Business Risk Management

Program executive officers and managers employ several business practices to manage risk. Among these, knowledge-based acquisition deserves attention, in part because it offers significant promise. According to the Defense Acquisition University, knowledge-based acquisition

> is a management approach which requires adequate knowledge at critical junctures (i.e., knowledge points) throughout the acquisition process to make informed decisions. DoD Directive 5000.1 calls for sufficient knowledge to reduce the risk associated with program initiation, system demonstration, and full-rate production. DoD Instruction 5000.2 provides a partial listing of the types of knowledge, based on demonstrated accomplishments, that enable accurate assessments of technology and design maturity and production readiness.

> Implicit in this approach is the need to conduct the activities that capture relevant, product development knowledge. And that might mean additional time and dollars. However, knowledge provides the decision maker with higher degrees of certainty, and enables the program manager to deliver timely, affordable, quality products.[20]

DoD strives to adhere to knowledge-based acquisition, and the government's acquisition systems and practices reflect this approach under the Rapid Equipping Force. However, it has found it difficult to hew to the principles involved while at the same time responding promptly to urgent operational needs generated by ongoing combat operations. Program managers with whom we spoke seemed to appreciate the benefits of knowledge-based acquisition and sought to adhere to best practices associated with knowledge-based acquisition in cir-

---

[20] Defense Acquisition University, "Knowledge-Based Acquisition," web page, June 21, 2004.

cumstances in which their programs are developing deliberately and not responding to crisis requirements.[21]

The JLTV program employs a technique that is closely related to knowledge-based acquisition called "knowledge point reviews."[22] Knowledge point reviews begin in the technology development phase to

> help the program frame which technologies were within performance reach, while trying to avoid high future JLTV life cycle and sustainment costs. . . . This approach will feed the service's [subsequent] knowledge point reviews to conduct whole system trade studies to refine the engineering and manufacturing development (EMD) phase requirements.[23]

Incremental acquisition appeared as a defense acquisition executive initiative first reported in the defense acquisition transformation report to Congress in 2007. As then conceived, it was intended to prioritize "joint and transformational capabilities to be deployed quickly to the warfighter."[24] Its effects have been broader, however, providing program managers with a mechanism to manage business risk enabled by open architectures. Because they know they are deploying the vehicles in increments, program managers need not force risky (immature) technologies into the design. They can rely on proven technologies today, knowing that, as improvements become available, they can be integrated into the development of the next increment of vehicles manufactured. Incremental acquisition therefore allows program managers

---

[21] This tension between knowledge-based progress and the desire to satisfy warfighting requirements has been recognized by GAO. See U.S. Government Accountability Office, *Defense Acquisitions: Opportunities for the Army to Position Its Ground Force Modernization Efforts for Success*, Washington, D.C., GAO-10-493T, March 10, 2010b.

[22] See Ashley John, "PEO CS&CSS JLTV Program Receives Top 5 DoD Program Award," *Accelerate*, Summer 2010.

[23] John, 2010, pp. 70–71.

[24] See Office of the Secretary of Defense, *Defense Acquisition Transformation: Report to Congress, John Warner National Defense Authorization Act, Fiscal Year 2007, Section 804*, July 2007, p. 26.

to maintain the basic architecture of the vehicle while incorporating upgrades to specific components as they become available, avoiding risky technologies and the cost growth and schedule delays they typically precipitate in the process.

Observations Concerning Operational and Program Requirements

Operational requirements tend to be transitory, reflecting DoD's thinking at the time about expected enemies, weather, terrain, and the best forces and operating practices with which to confront them. Operational requirements tend to change as reality diverges from the anticipated circumstances. Program requirements change more slowly to avoid the disruption of engineering and manufacturing processes, except where necessary. Necessary change can be introduced by a new, significant operational requirement or by the engineering change proposal process, invoked to address a program technical issue that has arisen. Given the long lives of Army and Marine Corps vehicles, it is highly likely that they will have to be modified at some point during their time in service, especially in light of modern adversaries' ability to adapt and present new challenges. Thus, the need for modifications is not, by itself, symptomatic of shortcomings in vehicle design and development.

As this chapter has illustrated, program delays can occur for a variety of reasons. The services have adopted a series of practices to reduce technical risk that could introduce time delays. They have also embraced business practices intended to reduce program risk, which serve to control both schedule delays and cost growth.

CHAPTER THREE

Vehicle Requirements and Gaps

Introduction

In Chapter Two, we discussed requirements, risk, and ways in which DoD is addressing some of these issues through better management practices. In this chapter, we sharpen our focus from general definitions and the management of service vehicle fleets to examine "capability gaps" and what they mean for DoD's ability to field adequate fleets of combat and tactical wheeled vehicles. As part of JCIDS, a capability gap analysis identifies capability gaps as the difference between a military *unit's* achieved performance and the tasks, conditions, and standards defined for it.[1] Note that the analysis is based not on the capabilities of any given system but on those of military units.

Getting from Capabilities to Requirements

The process of moving from capabilities (and capability gaps) to requirements is complex. Figure 3.1 summarizes the principal steps. As the figure indicates, a number of assessments are necessary to turn desired capabilities or operational requirements into program require-

---

[1] For more detail, see WBB Consulting, *Joint Capabilities Integration and Development System (JCIDS) Documents*, 2010. Some requirements emerge as straightforward performance criteria. Thus, a protection requirement may be expressed in terms of the caliber of ammunition it must protect against, e.g., "able to defeat 7.62 mm ammunition fired at the front quarter of the vehicle within 300 meters." Other times, requirements take a metaphorical cast, e.g., "as capable as a Bradley."

Figure 3.1
Principal Steps: Capabilities to Requirements

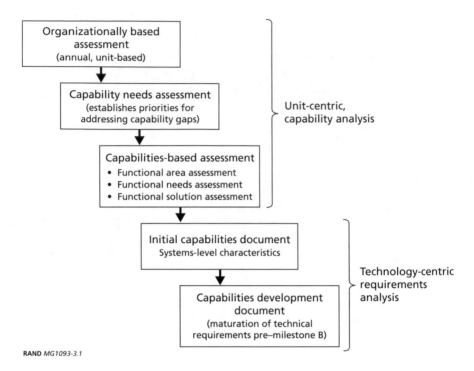

ments. The number of steps is driven, in part, by the fact that capabilities are often defined in terms of what a given type of unit (rather than a vehicle) must be able to do and are often expressed qualitatively, while program requirements must be expressed quantitatively. The process begins with the annual organizationally based assessment, which considers a specific type of unit (e.g., a Stryker battalion), its capabilities, and its capability gaps. Next, the capability needs assessment considers the capability gaps that have been identified for the unit and sets priorities for addressing them. At this point, the analysis has not settled on a materiel solution; the entire range of DOTMLPF receives consideration as potential parts of the remedy. In the course of the capabilities-based assessment, potential solutions are considered, including materiel solutions (i.e., new equipment or modifications to current equipment). These deliberations lead to the initial capabilities document and the systems-level characteristics that the materiel solution must deliver. The technology development phase of the program then defines the technical parameters necessary to satisfy the capability requirements by milestone B.

Achieving balance among performance criteria often involves making trade-offs. If trade-offs are not carefully managed, one design criterion (e.g., protection) can unduly dominate and distort the rest of the design, leading to vehicles ill suited for actual deployment. Many experts view the GCV designs that approached 70 tons as cases of runaway requirements. Satisfying all requirements is not synonymous with a good solution; final vehicle requirements represent a set of trades informed by assumptions about the enemy, weather, and terrain likely to confront U.S. forces at the time the requirements were approved.

The remainder of this chapter describes the requirements for the vehicles examined in the course of this study and concludes with a discussion on what their collective history has to say about the services' ability to close capability gaps and satisfy performance requirements.[2] As noted in Chapter One, this study was not an assessment of the individual vehicles but rather uses them as exemplars of particular classes of vehicles or examples of vehicles in a particular stage of development to help draw conclusions about the state of DoD's combat and tactical wheeled vehicle fleet and the department's ability to field an adequate one.

Ground Combat Vehicle

The GCV is envisioned to replace the Bradley fighting vehicle, which has been the Army's IFV for 30 years but no longer meets the requirements of current and future Army strategic, operational, and tactical fighting concepts. In testimony before the U.S. House of Representatives, Army officials succinctly stated that the GCV should have the "under-belly protection offered by MRAP, the off-road mobility and

---

[2] All technical references that correspond to specific vehicles come from their requirements documents, unless otherwise specified.

side protection of the Bradley Fighting Vehicle, and the urban and operational mobility of the Stryker."[3] In addition, the Army desires that the GCV transport and deploy nine-man squads as complete units, along with an undetermined number of crew members, while hosting advanced network capabilities.

On August 25, 2010, the Army canceled the original RFP issued in February of that year.[4] It released a revised RFP in November 2010. The new RFP seeks to minimize technology integration risks and requires that the vehicle be delivered within seven years of the contract award. According to the new RFP, the program intent is to "deliver an affordable and effective infantry fighting vehicle in seven years by utilizing mature technologies and focusing on system design that integrates all subsystems and components."[5] The Army's Chief of Staff, GEN George Casey, has further stated that the GCV will not be a "super-heavyweight."[6] In addition to minimizing risk and maintaining the program schedule, another important element for the revised GCV RFP, and likely all DoD programs going forward, is affordability.[7] The current RFP provides a "cost target [of] $9,000,000 to $10,500,000 per unit" in fiscal year (FY) 2010 constant dollars.[8]

---

[3] LTG Robert P. Lennox, LTG William N. Phillips, and David M. Markowitz, U.S. Army, "On Army Acquisition and Modernization Programs," testimony before the House Committee on Armed Services, Subcommittee on Air and Land Forces, March 10, 2010.

[4] One reason suggested for the cancellation was that the GCV was overly ambitious and had too many requirements and capabilities, which would have likely resulted in costly overruns and schedule delays. See Kate Brannen, "U.S. Army's GCV Delay: Lesson Unlearned," *DefenseNews*, August 27, 2010c.

[5] U.S. Army Contracting Command, Warren, Mich., Solicitation Number W56HZV-11-R-0001, November 30, 2010, p. 3.

[6] See Matthew Cox, "Casey: Make Ground Combat Vehicle Lighter," *Army Times*, June 21, 2010.

[7] C. Todd Lopez, "GCV Must Be Safe, Affordable, Full-Spectrum Capable," Army News Service, October 4, 2010b; Kris Osborn, "Army Leaders Brief Industry on Ground Combat Vehicles," Army News Service, October 5, 2010.

[8] U.S. Army Contracting Command, 2010, p. 4.

According to the November 2010 RFP, the program is driven to achieve four primary imperatives referred to as the "Big Four." These "Big Four" imperatives are defined as . . . Force protection . . . achieving a threshold level of protection for all occupants against the threats identified (in the classified threat annex); capacity is a vehicle crew and a fully-equipped nine soldier infantry squad; full spectrum is a versatile platform able to adapt and/or enhance capabilities through configuration change of armor and network while providing for growth over time in terms of size, weight, power, and cooling; and timing is the delivery and acceptance of the first production GCV IFV vehicle within seven years of the [technology development] phase contract award.[9]

Although the first RFP was canceled, it provides some insights into the issues examined in this research.[10] As the first combat vehicle to be designed from scratch in the era of IEDs, the original capability development document (CDD) emphasized the many system performance attributes associated with protection requirements generated by the IED threat. For example, the original RFP stipulated that the GCV would provide integral protection from life-threatening incapacitation from the various blast, shock, fragment, and acceleration effects of an attack. The protection will be scalable, with tailorable packages depending on mission requirements. The packages will be designed for application in the field by two-person crews with organic, brigade-level assets. Fire detection and suppression/extinguishing systems will be included.[11]

The mobility of the GCV will require that it remain capable of traversing the steep hills, valleys, and infrastructure typical in cross-country and urban terrain, as outlined in operational mode summaries and mission profiles and in the Army Test Command mobility

---

[9] U.S. Army Contracting Command, 2010, p. 4.

[10] While the requirements in the remainder of this section will likely change, they are provided here as an illustration of the capabilities that the Army at one point published as those it would like to have. In this sense, they illustrate important points.

[11] U.S. Army, Program Executive Office, Integration, *Technology Development Strategy, Ground Combat Vehicle*, draft, April 9, 2010, para. 2.3.

test requirements, while meeting survivability and force protection requirements.[12]

The GCV will be network-ready and able to exchange data in a secure manner that is compliant with DoD enterprise architecture and the net-centric data and services strategy while meeting the Global Information Grid technical, information assurance, and network supportability requirements. The vehicle will also have the capability to carry and control unmanned systems through onboard crew displays.

The GCV must retain the ability to perform its primary mission in a degraded mode and be operationally available at least 93.5 percent (threshold) or 96.8 percent (objective) of the time at full combat configuration when measured continuously over three-day missions. The GCV must also meet strenuous energy-efficiency criteria.

The GCV will be capable of defending itself against enemy troops and platforms at levels similar to the Bradley. This means that the GCV will achieve lethality overmatch against targets outlined in current threat assessments. These criteria will require a 95-percent probability of achieving desired effects with no more than two engagements and, ideally, with one engagement.

These requirements make clear that there are real trade-offs in translating operational requirements to program requirements that can be achieved within the cost and time constraints anticipated for this program. The November 2010 RFP includes three tiers of requirements[13] and

> offerors are encouraged to perform trade-off analysis within the available trade space of the GCV IFV requirement, defined as the Tier two and Tier three requirements . . . to provide an affordable, capable vehicle within seven years of [the technology development] phase contract award.[14]

---

[12] "U.S. Army, Program Executive Office, Integration, 2010, para. 2.1.3.

[13] Details for each tier are included in attachment 026 to the RFP.

[14] U.S. Army Contracting Command, 2010, p. 4.

Thus, the new RFP leaves the examination of the trade space and the feasibility of specific trade-offs to contractors. It is not clear that the Army has studied the feasibility of trade-offs for itself or has a firm sense of the trade space and its potential.

In addition, maintaining the SWaP-C reserve during the development process will be difficult and may place extra burdens on the process because of the uncertainties attending estimates of future weight, power, and cooling claims, but it may pay off down the line insofar as the reserves are sufficient.

Joint Light Tactical Vehicle

The JLTV is the replacement for the current high-mobility multipurpose wheeled vehicle (HMMWV). The original HMMWV design has been extensively modified to increase protection for its occupants at the expense of mobility and performance. The JLTV is envisioned to return the performance and payload lost through these modifications to the original HMMWV design. In short, the JLTV is designed to be a light tactical vehicle that will withstand IED attacks, maneuver easily through complex terrain, and be air-transported by a CH-47 helicopter.[15]

On September 19, 2007, John Young, then–Under Secretary of Defense for Acquisition, Technology, and Logistics, released a memo directing the military services and agencies to "formulate all pending and future programs with acquisition strategies and funding that provide for two or more competing teams producing prototypes through Milestone (MS) B." The JLTV program uses this "competitive prototyping" approach in the major, high-tech, and integration risk components of the vehicle before procurement to reduce those risks, gain knowledge, improve the designs, and assess the capabilities of the proposed manufacturing processes. BAE Systems, Lockheed Martin, and General Tactical Vehicles (a consortium of AM General and General

---

[15] Headquarters, U.S. Department of the Army, Army Requirements Oversight Council, "JLTV Family of Vehicles," briefing, March 28, 2007.

Dynamics Land Systems) have submitted vehicles for testing by Army and Marine Corps evaluators. After testing, contestants in the next phase, EMD, will be selected in an open competition. They may or may not be among the three that provided the original prototype vehicles. The services hope that such competitive prototyping will reduce the cost of the vehicle to the government.

The cost of a JLTV is expected to significantly exceed the cost of recapitalizing an HMMWV, and the assembly line for new, up-armored HMMWVs is closed. With more than 150,000 HMMWVs in the force needing eventual replacement or recapitalization, the cost per vehicle will continue to be an important consideration. In addition, differing joint mission requirements affect the height, weight, and other characteristics of the vehicle. The JLTV has armor protection and power for networking and sensors: capabilities in response to service requirements that affect performance and cost. The presence of significant numbers of MRAPs and M-ATVs in the force may reduce the need for other, robustly protected vehicles, such as the JLTV.[16]

As outlined in the JLTV CDD, the JLTV must have the ride quality, comfort, and safety to reduce passenger and crew fatigue and injury; allow passengers to perform critical missions during transit; and be able to fight on arrival. To this end, the JLTV should achieve or exceed by 10 percent (threshold) or 25 percent (objective) the rated mobility prescribed by the North Atlantic Treaty Organization (NATO) Reference Mobility Model.[17]

According to capabilities documents, JLTV designs must be consistent with intertheater strategic deployment and intratheater operational maneuver expectations by meeting the demanding air-transportability weight and size requirements for fixed-wing (C-130)

---

[16] As noted earlier, requirements are driven by perceptions of future conflict environments. The requirements for the JLTV are driven by threat environments that place a premium on its expanded capabilities. Should alternative scenarios, such as those that provide relatively secure rear areas (for which HMMWVs were designed), need to be planned for, the overall requirement for JLTVs would decrease.

[17] North Atlantic Treaty Organization, *NATO Reference Mobility Model*, Brussels, RTO-TR-AVT-107, 2002.

and rotary-wing (CH-47) aircraft.[18] The JLTV should meet the rail transportability requirements of the American Association of Railroads, Gabarit International de Chargement clearance diagrams, and the DoD Rail Clearance Diagram. It should be highway-transportable using both military and civilian transport and sea-transportable via existing strategic or intertheater shipping.

Joint and service command-and-control systems, as well as intelligence, surveillance, and reconnaissance (ISR) equipment must be seamlessly integrated and powered within the JLTV and meet mandated DoD Information Technology Standards Registry key interface parameters and profiles. It should host net-centric operations and Warfare Reference Model enterprise services, allowing assured and operationally effective information exchange.

Essential protection to mounted personnel is to be provided by inherent and supplemental armor exceeding the protection of an up-armored HMMWV in all threat domains and, eventually, meeting Interim Armor Classification level 4 in all categories.

In terms of protection, the JLTV should maintain structural integrity in a rollover, with a crush-resistant roof structure capable of supporting 150 percent of the ground vehicle's weight. For payload, the JLTV must be capable of effectively transporting payloads, weapons, and mounts in accordance with functional concepts. It should be operationally available 95 percent of the time and meet specified service-level maintenance ratios.

The JLTV reflects the challenges associated with efforts to build a vehicle suitable to different operational requirements and the needs of several services. Marine Corps imperatives call for air-transportability and vehicle size and weight consistent that are with its expeditionary mission and the storage space available for them aboard U.S. Navy ships. Army requirements, in contrast, emphasize protection. Thus, in satisfying the Army desire for protection, designers may invariably produce a prototype ill suited for Marine Corps applications. Otherwise,

---

[18] GAO reports that, should the JLTV fail to be air-transportable by CH-47F, the Marine Corps may back out of the program. See GAO, 2010e.

if designers honor Marine Corps preferences for size and weight, the resulting vehicle may reflect capability gaps from an Army perspective.

In circumstances in which satisfying requirements for one service undermines a vehicle's utility to another service, it may become necessary to adjust the role of the vehicle in question. For example, if the JLTV has to be light enough to meet Marine Corps requirements, then the Army might choose to substitute a mix of MRAPs and M-ATVs for JLTVs in roles in which protection is at a premium. The broader point is that joint programs should not be created around vehicles when the military services have widely different requirements and a need for different designs.

The most significant challenge to this program appears to be fiscal rather than physical (assuming that weight can be kept low enough given protection requirements). In particular, the JLTV is far more expensive than the HMMWV that it will, in part, replace. (It is less expensive than the MRAPs it will also replace, however.) A large portion (approximately 50 percent) of DoD's overall tactical wheeled vehicle fleet consists of light vehicles (e.g., there are more than 120,000 HMMWVs in the Army fleet alone).[19] If the vehicle meets requirements but DoD cannot afford to replace all other light vehicles with it, then increased operational risk will have to be accepted.

Expeditionary Fighting Vehicle

The Marine Corps EFV represents an attempt to develop a vehicle that can assault a defended beach at a reasonable distance from supporting vessels and move inland while protecting and supporting the marines inside. This is a heroic task, and the program has undergone three re-

---

[19] Headquarters, U.S. Department of the Army, *Army Truck Program (Tactical Wheeled Vehicle Acquisition Strategy) Report to the Congress*, June 2010a, reports that the Army currently has approximately 160,000 HMMWVs of various sorts on hand or due in. The Army projects that it will reduce this number to 85,000 by 2025. The JLTV buy was not reported in that document (p. 5). GAO, 2010e, states that the base system costs are $445,000 for the M-ATV, $186,000 for up-armored HMMWVs, and a projected $306,000–$332,000 for the JLTV.

baselines since entering the first system development and demonstration (SDD) phase in 2000. The first re-baseline occurred in November 2002, then again in March 2003, and most recently in March 2005. In 2006, the Marine Corps produced the *Capability Production Document (CPD) for Expeditionary Fighting Vehicle (EFV) Increment: Single Step to Capability Low Rate Initial Production (LRIP)*, which listed the eight key performance parameters (KPPs), plus threshold and objective values, for 279 additional attributes.[20]

This same year, GAO produced a report outlining the troubled history and future challenges for the EFV.[21] In early 2007, the EFV program was restructured (again), this time triggering a Nunn-McCurdy review. A memo from the Joint Requirements Oversight Council concerning the EFV Nunn-McCurdy certification recertified each of the KPPs except that for high water speed.[22] The high-water-speed KPP (Section 6.2.1.1) from the 2006 CPD stated that the EFV (personnel variant) had to achieve "an average high water speed of 20 knots [a KPP threshold], 25 knots . . . [objective], for at least one continuous hour, while in combat-loaded condition, in water with a significant wave height (SWH) of 0.91 meters (3 feet)." According to the May 2007 memo, the water-mobility KPP was relaxed such that the EFV "shall be capable of transiting 25 nautical miles in seas with a significant wave height of two feet at an average of 20 knots (Threshold); 25 knots (Objective)."[23]

In September 2007, the Marine Corps Combat Development Command issued a memo specifying further modifications to the

---

[20] U.S. Marine Corps, "Expeditionary Fighting Vehicle (EFV) Capability Production Document (CPD)," April 13, 2006, Table 6.2.

[21] U.S. Government Accountability Office, *Defense Acquisitions: The Expeditionary Fighting Vehicle Encountered Difficulties in Design Demonstration and Faces Future Risks*, Washington, D.C., GAO-06-349, May 1, 2006.

[22] U.S. Joint Chiefs of Staff, "Expeditionary Fighting Vehicle Nunn-McCurdy Certification," Joint Requirements Oversight Council Memorandum 108-07, May 8, 2007.

[23] U.S. Joint Chiefs of Staff, 2007.

EFV requirements.[24] In addition to the modification to the high-water-speed KPP, it also made changes to two non-KPP requirements. The first reduced the on-land operational range after a 25-nm swim from 200 miles on hard-surface roads to 100 miles. The second modification eliminated the requirement for smoke obscuration for both force protection and survivability.

Reliability requirements, too, have been reduced over the history of the EFV. The original operational requirements document (ORD) for the EFV[25] stipulated that the mean time between operational mission failures have a threshold value of 70 hours and an objective value of 95 hours.[26] According to a GAO report,[27] in January 2005 the Joint Requirements Oversight Council approved reducing the mean time between operational mission failures from the original threshold and objective values to 43.5 hours (threshold) and 56 hours (objective).[28]

The EFV successfully met all the then-current requirements for cost, schedule, and performance as part of a 2008 critical design review but, according to a 2010 GAO report, there remains a "significant risk associated with achieving required reliability," as demonstrated in the initial operational test and evaluation.[29]

---

[24] U.S. Marine Corps, Deputy Commandant for Combat Development and Integration, "Expeditionary Fighting Vehicle (EFV) Capability Production Document (CPD); Change 1," September 26, 2007.

[25] In 2003, the Advanced Amphibious Assault Vehicle Program was renamed the Expeditionary Fighting Vehicle Program. See H. Keeter, "Marine Corps' AAAV to Be Renamed 'Expeditionary Fighting Vehicle' (Advanced Amphibious Assault Vehicle)," *Defense Daily*, July 1, 2003.

[26] U.S. Marine Corps, "Operational Requirements Document for Advanced Amphibious Assault Vehicle (AAAV) ACAT I-D Prepared for Milestone II Decision," No. MOB 22.1, September 13, 2000.

[27] U.S. Government Accountability Office, 2006.

[28] U.S. Marine Corps, 2006.

[29] U.S. Government Accountability Office, *Expeditionary Fighting Vehicle (EFV) Program Faces Cost, Schedule, and Performance Risks: Briefing for the Subcommittee on Defense, Committee on Appropriations, House of Representatives*, Washington, D.C., GAO-10-758R, July 2, 2010d.

In 2009, the Marine Corps Combat Development Command, responding to congressional concerns about the vulnerability of the flat hull design to underbody IED attacks, made an additional change to the 2006 EFV CPD.[30] According to the memo and supporting information, several alternatives for improving the underbody protection were evaluated, including integration of an external V-shaped hull, an internal V-shaped structure, and the addition of underbody appliqué armor. The appliqué armor was determined to be the most cost-effective solution that would have least impact on the current schedule. However, appliqué armor could be attached only once the vehicle was on land and prevented water mobility; it also required additional lift to transport the appliqué armor to the area of operations.

In a recent memo on achieving greater efficiency and productivity in defense spending, Under Secretary Carter stated the need for making affordability a requirement. Specifically, the memo suggests that programs preparing for milestone A should include an affordability target "to be treated by the program manager (PM) like a Key Performance Parameter."[31] Beginning in 2000, the program unit cost and average procurement costs were estimated to be $8.8 million and $7.2 million (in FY 2011 dollars), respectively. The President's 2011 budget estimates that program unit costs and average procurement costs have increased to $24.3 million and $18.4 million, respectively. While the estimated development and procurement costs have increased only 2.5 percent and 3.5 percent since the 2007 Nunn-McCurdy breach and restructure, the 2010 GAO report suggests that additional risks may further increase costs as the program goes forward.[32]

The history of the EFV reflects the vehicle's importance in future Marine Corps amphibious assault concepts (indeed, in maintaining amphibious assault as a viable undertaking against very capable future adversaries), and this importance explains, in part, why the service

---

[30] U.S. Marine Corps, Deputy Commandant for Combat Development and Integration, "Expeditionary Fighting Vehicle (EFV) Capability Production Document (CPD); Change 2," May 20, 2009.

[31] Carter, 2010c.

[32] U.S. Government Accountability Office, 2010d.

accepted the increased costs associated with technology, performance, and reliability risk as the program unfolded. Vehicle programs that are less central to the future of the Marine Corps almost certainly would have been canceled. The history of the EFV also suggests the limitations bounding the prospects for improving performance and reliability. In both instances, requirements had to be relaxed in the face of insurmountable technical challenges. Finally, the EFV's history reflects how unit costs can grow as a result of developmental issues, to the point that the service must reduce the number of vehicles it ultimately seeks to acquire to fit within its available funding.

The EFV also highlights the tensions between ambitious operational and program requirements and the challenges of pushing the limits of technology and engineering to meet those requirements. In addition to the technical and traditional engineering challenges of getting a vehicle of that weight to travel quickly over long distances by water and arrive ready to fight (both crew and machine), systems engineering challenges and the workforce to manage large, complex programs are thought to have been major problems early in the program (though experts outside the Marine Corps told us that these workforce concerns have been overcome).[33] In particular, adequate systems engineering up front to ensure that complex programs are "born healthy" could have helped identify some of the technical and financial challenges that would later put the EFV in a bad light.

These technical and engineering challenges, in turn, present significant challenges from the perspective of funding. In particular, our consultations with experienced program managers and other senior personnel in the acquisition system indicate that sufficient and stable funding is a critical issue for system success. Yet, in this case, the funding challenges stem from low cost estimates and growing costs—an important element in ensuring that funding is adequate and one that, according to those we have consulted, is an ongoing challenge.

---

[33] Statements from the project's SME workshop, July 2010. In particular, the Marine Corps does not often build large, complex programs and may have experienced a steep learning curve with the EFV—a learning curve that it has apparently surmounted, according to workshop participants from the Office of Naval Research.

Medium Tactical Vehicle Replacement

The original mission needs statement for the MTVR was published in 1992.[34] The vehicle was intended to address a perceived gap in the current tactical wheeled vehicle fleet. Specifically, according to the statement, the tactical fleet lacked the mobility and on-road/off-road payload capacity to support Marine Air-Ground Task Force combat support and combat service support units as outlined in the task force's June 28, 1991, master plan.[35] The corresponding ORD for the MTVR was published in 1994.[36] Five variants were identified as requirements: a 6×6 cargo truck, a 6×6 extended-wheelbase cargo truck, a 6×6 dump truck, a 6x6 wrecker truck, and a 6×6 tractor truck. These five variants were needed to perform the following functions: artillery prime movement, transport of troops and equipment, bulk cargo and ammunition hauling, water and fuel container transport, shelter transport, dump capability, and wrecker/retriever activities. It is worth noting that, outside of the requirement for electronics and communication capabilities listed here, the ORD does not anticipate any additional computer resources.

Since the original ORD was released in 1994, there have been relatively few changes to the stated operational requirements. However, in 1997, a memo from the Marine Corps Combat Development Command deleted the requirement for the add-on armor kit (item 19). Oshkosh Corporation began producing MTVR vehicles for the Marine Corps in 1999.

Currently, the MTVR consumes 50 percent of all fuels used by Marine Corps vehicles on the battlefield,[37] which makes its fuel con-

---

[34] U.S. Marine Corps Combat Development Command, "Mission Need Statement, Medium Tactical Vehicle Replacement," No. MOB 211.4.2A, March 1992.

[35] U.S. Marine Corps Combat Development Command, 1992, sec. 3.

[36] U.S. Marine Corps Combat Development Command, "Operational Requirements Document (ORD) for the Medium Tactical Vehicle Replacement," No. MOB 211.4.2A, January 1994.

[37] Glenn W. Goodman, Jr., "S&T and Cost Estimating," in Scott R. Gourley, *Nearly Four Years in Operation: Program Executive Officer Land Systems Marine Corps Looks Ahead to the Future*, Quantico, Va.: Program Executive Office, Land Systems, U.S. Marine Corps, 2010.

sumption a major consideration—an issue that has implications for how vehicles are designed and purchased. In particular, because funds are allocated annually and within categories, operations and maintenance costs (which constitute a major element of a system's life-cycle usage costs) may not be fully factored into development and acquisition decisions. More comprehensive costing models might lead to decisions to purchase vehicles that are more expensive to develop and buy but less expensive to develop, buy, and operate over their life spans.[38]

In modern military operations, logistics vehicles are used through every phase and appear in every part of the battlefield; therefore, they must be prepared for the full suite of conditions and threats that may confront them. According to June 2006 testimony by Maj. Gen. William Catto before the House Armed Services Committee, the Marine Corps initiated an engineering change proposal with Oshkosh Corporation in November 2003 for the creation of the MTVR armor system.[39] The armor system would be a permanent modification to the MTVR and is designed to last the lifetime of the vehicle (21 years). According to Major General Catto's testimony, the MTVR armor system "is capable of withstanding small arms fire, IEDs, and mine blasts up to 12 pounds. It consists of metal/composite panel armor, with separate cab and troop compartment kits, dependent upon cargo or personnel variants of the MTVR." In September 2004, the Marine Corps ordered 796 MTVR armor systems from Oshkosh Corporation. By January 2006, the first system was fielded with a Marine expeditionary unit. As of May 2006, 874 MTVRs had been modified with the armor system, in a span of less than two years from the initial engineering change proposal to completion.

The MTVR is an example of the minimalist approach to repurposing commercial vehicles. As originally conceived, very little dis-

---

[38] U.S. Marine Corps representatives at our July 2010 workshop strongly emphasized this as a critical issue. Changes under way in how DoD calculates cost estimates, in response to Under Secretary Carter's memorandum on the subject (Carter, 2010c), may result in advances on this front.

[39] Maj. Gen. William Catto, U.S. Marine Corps, testimony before the House Committee on Armed Services, Subcommittee on Tactical Air and Land Forces, on Marine Corps force protection efforts, June 15, 2006.

tinguished it from its civilian counterparts. It was not developed as a network-ready tactical vehicle with crew protection. The assumptions that drove its development centered on mobility. Later in its life, however, MTVR came to face threats that prompted the Marine Corps to revise its assumptions about the vehicle's requirements and retrofit it with armor protection. In addition to lightweight armor, future improvements and upgrades to the MTVR's suspension, automatic fire suppression systems, common displays, and vetronics are planned.

MTVR is an example of a successful program that leveraged commercial technology and adapted quickly to meet the operational requirements of conflict scenarios for which it was not designed.

Heavy Expanded Mobility Tactical Truck

The HEMTT forms the core of the Army's future heavy tactical wheeled vehicle fleet and is designed and built solely by Oshkosh Corporation, which has delivered in excess of 20,000 HEMTTs in many versions.[40] Some of the variants of the HEMTT are the M977/997/985 cargo, the M978 fueler, the M982/983 tractor, and the M984 wrecker/tow vehicle. The M1074/75 palletized load systems (PLSs) and PLS trailers automate the loading and unloading of the items they carry. The HEMTT also transports air defense systems, as well as firefighting, construction cranes, cement mixers, and other service equipment. The Army's heavy armored vehicles, when not engaged in combat operations, are carried on flatbed M1000/1070 Heavy Equipment Transporters, versions of the HEMTT.

The requirements placed on the HEMTT fleet are numerous. The trailers must have a gross vehicle weight rating of 40,000 lb. The cargo variant must have a payload of 22,000 lb. The tractors must be mobile over a range of terrains, from firm ground to soft soil, sand, mud, and snow. Recent requirements specify space and power for movement-tracking systems. The PLS variant must be able to load and unload an

---

[40] Louis Anulare, Program Executive Office, Combat Support and Combat Service Support, "NDAA Section 222," Washington, D.C., briefing, May 12, 2010.

11-ton payload with only one soldier. The refueling variant has a payload of 2,500 gallons. The wrecker variant must be able to recover and evacuate all types of U.S. Army wheeled vehicles. HEMTTs should have a mission reliability of 1,500 mean miles between mission failures and a system reliability of 500 mean miles between system failures.

The HEMTT fleet is split between the A2 version and the newer A4 version. The A4 production variants are a significant improvement over the A2 variant, and all the A2s in the fleet are in the process of being recapitalized into A4 variants at a rate of 590 per year from FY 2011 through FY 2027. The A4s feature a 500-hp Caterpillar C-15 engine, a 600-hp Allison 4500 SP/5-speed automatic transmission, power train and suspension upgrades, and improvements to the cab with improved climate control. The U.S. Army's Long Term Armor Strategy (LTAS) drives many of the cab improvements. The base armor of the LTAS-A kit on the cab can be augmented with standardized, bolt-on armor from the LTAS-B kit. The 12,000+ HEMTT versions prior to the A4 have no armor upgrade capability using a B-kit.[41] Integrated mounting points built into the cab of the A4 allow fast installation of the B-kit, the protected gunner position, and the machine-gun mount on the cab roof.

According to Army participants in the project's workshop, the Army leverages commercial truck developments for use in military trucks, but it is losing market share and thus its ability to drive improvements inexpensively. This is especially true for engines and transmissions.

The HEMTT also illustrates potential changes in acquisition system rules that would add efficiency. Since the HEMTT is similar to commercial trucks and has been in the fleet for a long time, it is a mature platform about which much is known. Modifications (for example, to upgrade its transmission, as was done for the A4 model) add to the overall expense of the program and may trigger ACAT (acquisition category) I-D oversight requirements. However, since both the system and the new transmission are mature technologies and there

---

[41] LTC Allen Johnson, "PM Heavy Tactical Vehicle," briefing presented at the 2010 Tactical Wheeled Vehicles Conference, Monterey, Calif., February 7–9, 2010.

is very little technical risk from a systems integration perspective, this level of oversight appears to be uncalled for (though it is mandated under current rules by the overall cost of the program).[42] A revised approach to oversight requirements and program management that is based on risk rather than total cost might save such programs considerable time and expense. If such an approach were adopted, it would be relatively easy to, for example, install proven, mature transmissions in such trucks rather than creating a new increment to the program in order to upgrade the transmissions.[43]

If the MTVR represents the minimalist approach to repurposing commercial vehicles, the HEMTT represents the maximalist approach. The HEMTT has evolved to reflect the needs of the sustainment community, which it serves. In that role, the vehicles are sufficiently network-ready to support the community's inventory visibility system, the Military Tracking System. Some HEMTT models (e.g., PLS) include materiel-handling equipment to reduce crew workload. It has received protection upgrades as part of the LTAS. The HEMTT represents the best case or success story for a military program adapting and evolving its vehicles in response to changes in assumptions and the real-world circumstances in which these vehicles must operate.

M1A1/M1A2 Abrams Main Battle Tank

The purpose of a main battle tank, such as the Abrams, is to provide unmatched mobile firepower on the move with maneuverability and protection for the crew.[44] Firepower is provided by a 120-mm M256 smoothbore cannon, a coaxially mounted 7.62-mm M240 machine gun, and a 0.50-caliber M2HB commander's weapon. The hull and turret of the Abrams are protected by special advanced-composite

---

[42] Discussions with SMEs during the project's workshop, July 2010.

[43] An example provided by one of the participants in the project's workshop, July 2010.

[44] The M1 Abrams main battle tank is the first of the tier 2 systems examined in this monograph.

armor. The Abrams is powered by a 1,500-hp Lycoming Textron AGT gas turbine.

The M1 Abrams main battle tank weighs between 70 and 75 tons after evolving through seven major variants. The initial M1 version was introduced in 1980, the M1A1 in 1985, the M1A1 ED in 1999, and the M1A1 SA in 2006. These versions are all based on analog architectures. The M1A1 ED and M1A1 SA are part of the Abrams Integrated Management Program, which involves rebuilding older M1s to like-new condition. Models based on digital architectures include the M1A2 introduced in 1992, the M1A2 SEPv1 introduced in 1999, and the current-production M1A2 SEPv2, which was introduced in 2007. There are about 5,500 Abrams tanks now in the Army inventory, with many more in the Marine Corps and allied armed forces.

According to the 1994 M1A2 ORD, the maximum height of an M1 is 96 inches, and the maximum width is 144 inches.[45] The tank's maximum range is approximately 243 miles with the nuclear, biological, and chemical system on, and 256 miles with it off, at a maximum speed on flat, paved roads of 41.5 mph and cross-country of 30 mph. The M1A2 can accelerate from 0 to 20 mph in 7.2–7.5 seconds with its nuclear, biological, and chemical system off and 9 seconds with the system on. The mean miles before failure should be 320, with a maximum maintenance ratio of 1.25 maintenance labor hours per operating hour.

The M1A2 SEPv2 system enhancement package, according to civilian sources, included depleted-uranium armor, digital command-and-control architecture, digital color terrain maps, and new sensors.[46] The M1A2 SEPv2 has one noteworthy gap in its silent watch capability. The objective duration for this capability is 12 hours of operation when fully charged.[47] Silent watch requirements for the M1-series tank

---

[45] Headquarters, U.S. Department of the Army, Office of the Deputy Chief of Staff for Plans and Operations, Force Development, *Abrams Modernization Program M1A2, Operational Requirements Document*, Washington, D.C., January 1994.

[46] "M1 Abrams Main Battle Tank," MilitaryPeriscope.com, last updated May 1, 2009.

[47] Headquarters, U.S. Department of the Army, Office of the Deputy Chief of Staff for Plans and Operations, Force Development, 1994.

include sights operational, radios on silent listening mode, heater on if required, turret capable of traversing, and gun tube capable of elevating with the nuclear, biological, and chemical system on. The interim solution is a pack of six lead-acid batteries providing eight rather than 12 hours of silent watch capability when fully charged. A longer-term solution is an underarmor auxiliary power unit. This solution is required in the draft Abrams CDD as part of the Abrams modernization strategy. A waiver would be possible for the current M1 platforms.

The Abrams faces SWaP-C challenges as additional modifications are integrated into the vehicle. These modifications include those brought on by increased protection requirements, such as underbelly armor, a mine-resistant driver seat, and reactive armor tiles. Other improvements motivated by lessons learned from operations are the lower weight, shorter-recoil XM360E1 cannon, better target-acquisition sensors, a high-voltage integral generator, and an electric gun and turret drive. All these improvements require additional power, transmission, and suspension upgrades. The Army is developing draft initial capabilities documents, CDDs, and CPDs for these changes, but no additional funding has yet been identified.

The Abrams example illustrates how even the very best ground combat vehicles can require upgrades or additional capabilities as the circumstances in which it is employed change. The Abrams also demonstrates the difficulties that can arise when a vehicle's SWaP-C budget is small and accommodating new requirements for power generation or other capabilities can exceed the available SWaP-C space.

Paladin Integrated Management

PIM is a service life extension program. Our examination of it is limited to its automotive components. The system consists of the M109A6 Paladin self-propelled 155-mm howitzer and the M992A2 field artillery ammunition supply vehicle. The M109 series howitzer was first introduced in 1962 and today provides the Army's only fielded self-propelled artillery capability. Many of its components are old and obsolete. System weight has increased by 21 percent, while horsepower has increased by only 8 percent. Needless to say, many of the key PIM improvements are in automotive technologies.

Today's Paladin suspension and drive train do not allow growth in operational capability and survivability. Improvements are needed to the engine transmission, final drives, and suspension. The solution is to leverage Bradley and FCS non-line-of-sight cannon technologies. The suspension and track will be improved by new road arm stations, torsion bars, rotary dampers, and a 19.1-inch track. The power train will start with a new 600-hp engine. An HMPT 500-3ECB transmission will connect to the final drive, with power takeoff and a shift tower, along with brakes and steering. These components are common to the PIM and the Bradley and have TRLs between 6 and 9.

PIM demonstrates skill in adaptive engineering, moving an artillery system to new components when the original versions no longer support needed upgrades. PIM also suggests that there is virtue in having a multiplatform fleet that will allow something like a cannon system to migrate to more capable chassis components when necessity dictates. The PIM case also illustrates how the Army has sought to harvest useful concepts and technologies and apply them to today's vehicles—in this case, the suspension and power train components from the Bradley and the electric elevation and transverse drives from the non-line-of-sight cannon system.

Stryker

The Stryker is an eight-wheeled combat vehicle, purchased as an interim solution to the FCS while it was in production and because it is air-transportable in a C-130. The ten configurations of the Stryker are as follows: an infantry carrier vehicle, a mobile gun system, an ATGM vehicle, a commander's vehicle, a mortar carrier, a reconnaissance vehicle, an engineer squad vehicle, a nuclear-biological-chemical reconnaissance vehicle, a medical evacuation vehicle, and a fire support vehicle. Of the 4,443 authorized, there are 3,930 Strykers on hand or on order. The basic Stryker is C-130–transportable, with 14.5 mm of basic armor protection. Field kits add additional protection and weight, but if the vehicles are to be moved by a C-130, these kits must be air-transported separately. The Stryker is the primary vehicle for the nine Stryker brigade combat teams (SBCTs).

In November 2000, the interim armored vehicle (i.e., Stryker) contract was awarded. Delivery of new vehicles began in April 2002. In May 2003, initial operational test and evaluation began, and in November 2003, the Stryker entered operational service with the U.S. Army. Five years later, in August 2008, the Army was directed to mitigate deficiencies discovered while testing the Stryker's mobile gun system. Chief among these deficiencies were problems with the handling of ammunition and the reactive armor tiles. Improvements are part of the Stryker modernization strategy. On May 19, 2010, the Defense Acquisition Executive directed the Army to provide the overarching integrated product team with a complete overview of Stryker modernization efforts within 60 days. The Army requested a delay because of ongoing combat portfolio reviews as part of the combat vehicle modernization effort. The program manager was expected to brief Army leadership (as directed by the Vice Chief of Staff on September 10, 2010) and provide a full layout, with potential options, of the Stryker Modernization Program prior to January 2011. Following concurrence from Army staff on the path forward, the Army will reschedule the directed overarching integrated product team.[48]

According to the SBCT program manager, in 2010, the Stryker family of vehicles had capability gaps in four areas: protection, mobility (as modified with additional armor), lethality, and networking.[49] The protection capability gap centered on underbelly protection against IEDs and mines. A double-V hull upgrade for part of the Stryker fleet is under development and is currently undergoing testing.[50] The double-V hull, along with other modifications, may make the vehicle

---

[48] Program Manager, Stryker Brigade Combat Team, "Stryker Brigade Combat Team (SBCT) Modernization Strategy," briefing, September 22, 2010, slide 6.

[49] Program Manager, Stryker Brigade Combat Team, 2010, slide 6.

[50] David W. Duma, Principal Deputy Director, Operational Test and Evaluation, Office of the Secretary of Defense, testimony before the Senate Committee on Armed Services, Subcommittee on Airland, April 15, 2010.

less stable and affect its maintainability. In addition, protection against kinetic, chemical, and tandem warheads is required, as is protection for the vehicle occupants against rapidly accelerating advances in enemy weapons. The suspension will need to be upgraded to handle the increased weight that accompanies this level of protection.

Originally, the Stryker design emphasized lightness, speed, mobility, and air-transportability, but add-on protection kits and other modifications have affected its capability to support widely dispersed and continuous operations across a range of terrains. A major gap is the lack of mobility in restricted quarters. The Stryker also needs enhanced direct-fire capability against like threat systems and lacks sufficient space and power to accept mission equipment packages, such as non-lethal weapons.

In the area of network communication, the Stryker lacks adequate onboard power for the required sensor-to-shooter equipment to enable rapid situational awareness in all conditions and the ability to conduct cooperative engagements.[51] The Stryker, in general, lacks sufficient SWaP-C to support current needs and future required enhancements. Planned solutions to these deficiencies include a new 450-hp power pack; a larger, combined heating-cooling unit; and increased power generation and distribution capability.

Current plans anticipate modernizing the Stryker family of vehicles over the course of the next ten years,[52] including across SWaP-C (e.g., use of kits to provide greater ballistic protection, weight reductions to improve mobility, and electrical power-generation enhancements to reduce heat generated, produce larger power loads, and support silent watch). Modernization will be incremental. Technology development in increment II will begin in FY 2013, then move to EMD in FY 2015,

---

[51] According to a briefing by COL Robert W. Schumitz, Stryker Brigade Combat Team Project Manager, "PM Stryker Brigade Combat Team," briefing for the Program Executive Office, Ground Combat Systems, advance planning brief for industry panel, October 22, 2010b.

[52] Schumitz, 2010b.

culminating at milestone C in mid-FY 2018. The notional schedule anticipates new prototype vehicles concurrent with the EMD phase.[53]

The Stryker case exemplifies the challenges that can arise when a vehicle is procured under one set of assumptions and deployed under another. Originally a stopgap measure to maintain the Army's relevance in crisis-response operations prompted by events in the Balkans in 1999, and subsequently refined as an "interim" combat vehicle until FCS reached fruition, Strykers currently serve as the backbone of seven brigade combat teams and have become an integral part of the Army force structure. The Army has worked vigorously to adapt these vehicles to current operating conditions in Afghanistan and elsewhere, but capability gaps, as noted, persist. The lesson, however, is not just that gaps remain; rather, it is that the research, development, and acquisition communities took an existing LAV-25 with limited utility for the then-envisioned or current conflicts and turned it into a much more capable vehicle.

Observations on Closing Capability Gaps and Satisfying Performance Requirements

As the vehicle programs described in this chapter make clear, closing capability gaps and addressing performance requirements are difficult tasks. Part of the difficulty arises from the cycle of action-reaction between U.S. and enemy forces as they seek tactical advantages over each other. With tactical wheeled vehicles like the HEMTT, part of the difficulty lies in the fact that these vehicles are "repurposed" commercial trucks.[54] The more extensive the modifications necessary to close capability gaps or satisfy current performance expectations, the more expensive the work is likely to be. The challenge, however, is not limited to tactical wheeled vehicles.

---

[53] Schumitz, 2010b, slide 35.

[54] An observation by an engineer and branch chief at TRADOC during discussions with the project leaders, October 20, 2010.

All vehicles, as the examples in this chapter illustrate, must manage the competing demands of operational requirements for power, protection, and payload/performance that manifest as SWaP-C requirements. Sometimes, one of these considerations (e.g., protection) dominates the equation and suppresses the other design criteria, as in the case of the GCV. Sometimes, the appearance of new operational requirements can cause a new vehicle to evolve into a much different and more expensive vehicle than the one it replaces, as in the case of the JLTV, the successor to the HMMWV. The current emphasis on afford-ability places additional constraints on the services' ability to manage SWaP-C, develop materiel solutions to close capability gaps, and satisfy evolving performance requirements, as the PIM case demonstrates.[55] There are also instances in which a vehicle can play such a central role in a service's mission, as the EFV does in the Marine Corps, that the service will accept lengthy schedule delays, significant cost growth, and substantially revised performance criteria to preserve the capa-bilities that the vehicle provides. Finally, Stryker illustrates what can happen when circumstances surrounding a vehicle change dramati-cally. The Stryker began life as an interim vehicle until FCS reached fruition and emphasized air-transportability for crisis responsiveness over other design considerations. Subsequently, the FCS was canceled, which thrust the Stryker into a new, extended role as a major part of the Army force structure under circumstances substantially different from those anticipated when the vehicle was initially acquired. As a result, the Stryker has capability gaps in protection, mobility, lethality, and networking, despite vigorous Army efforts to adapt it for current operations and illustrates the challenges of forecasting requirements far into the future.[56]

---

[55] While affordability has always been an issue, Under Secretary Carter has recently placed additional emphasis on it. See Carter, 2010c.

[56] See Todd Lamb, Program Manager, Development, "Stryker Modernization Update (AUSA)," briefing presented at the meeting of the Association of the United States Army, October 6, 2009, p. 2. He states, "Current Space, Weight, and Power Capacity Shortfalls require upgrades to Stryker [family of vehicles]."

CHAPTER FOUR

Trade-Offs Among Vehicle Design Criteria to Improve Performance for Anticipated Circumstances

The process of designing any military vehicle involves trade-offs across the various vehicle design considerations. These elements include factors such as survivability and protection; mobility; power; munitions; armament; command, control, communication, intelligence, surveillance, and reconnaissance capabilities; maintenance; and deployment. This chapter highlights some of the observed trade-offs between various design considerations for the vehicles included in this study, as well as comments on emerging technologies and how they influence the design of future combat and tactical wheeled vehicles.

Weight Versus Mobility

Over the past decade, a trade-off that has been frequently discussed in the context of vehicle design is that between weight and mobility. One of the most significant sources of additional weight on ground vehicles is the increased armor used to improve vehicle survivability and protect its occupants. The additional armor is a direct response to the expanded use of (and improvements to) IEDs, explosively formed projectiles, and other similar devices against U.S. forces during operations in Iraq and Afghanistan. However, the added weight comes at a cost; it increases wear and damage to vehicles' suspensions, drives up maintenance requirements and fuel consumption, and limits mobility, particularly in difficult, off-road conditions.

A specific example of the trade-off between mobility and weight has been the addition of appliqué armor to the Marine Corps' EFV.

As mentioned in Chapter Three, the Marine Corps Combat Development Command altered the 2006 EFV CPD in response to congressional concerns about the vehicle's potential vulnerability to underbody IEDs to include add-on armor for additional protection. The use of appliqué armor was determined to be the most cost-effective solution that would least affect the current schedule. However, the appliqué armor also had a significant impact on the vehicle's mobility. Once the armor was attached, the EFV was limited to ground movement only and could not be used in its amphibious mode until after the appliqué armor was removed.

Another example of the trade-offs between weight and mobility is the JLTV. Viewed as the future Army's and Marine Corps' light tactical vehicle, the lightest of the three variants (the JLTV-A) weighs approximately 7.5 tons, which is three times heavier than a standard HMMWV (which is 5,200 lb, or 2.6 tons).[1] Much of the JLTV's added weight comes from requirements associated with improved protection and vehicle survivability, as well as integrating lessons learned from recent conflicts in Iraq and Afghanistan.[2] The added weight presents issues for both services; however, it is especially challenging for the Marine Corps because of its need for light, expeditionary vehicles that must be quickly transported to and around the battlefield. The weight of the JLTV may hamper the vehicle's ability to be transported to battlefields by air or sea.[3]

The trade-off between weight and mobility has been at the center of the recent debate concerning the proposed weight for the GCV. Some proposals have been estimated its weight to be approximately

---

[1] "HMMWV (High Mobility Multipurpose Wheeled Vehicle)," *U.S. Army Factfiles*, undated.

[2] Kate Brannen, "Efficiency Push Could Threaten U.S. JLTV," *DefenseNews*, October 7, 2010d.

[3] Andrew Feickert, *Joint Light Tactical Vehicle (JLTV): Background and Issues for Congress*, Washington, D.C.: Congressional Research Service, RS22942, September 17, 2010; Kate Brannen, "Mobility vs. Survivability: JLTV Could Suffer as U.S. Army, Marines Diverge," *DefenseNews*, June 7, 2010b.

70 tons.[4] This would put the vehicle at roughly twice the weight of the Bradley, which it is intended to replace, and almost as heavy as an Abrams M1 tank. The added weight is due largely to the armor needed to improve survivability and protection. According to the *2010 Army Modernization Strategy*, the GCV was intended to "provide a versatile range of capabilities, including the underbelly protection offered by MRAP, the off-road mobility and side protection of the Bradley Fighting Vehicle, and the urban and operational mobility of the Stryker."[5] However, the proposed weight to achieve the underbelly protection of an MRAP has revealed concerns about achieving the desired mobility. In particular, recent comments by GEN George Casey suggest that extremely heavy vehicles, like tanks, Bradleys, and others, are not practical in urban combat because of their size and lack of maneuverability. Currently, the revised GCV RFP that appeared in November 2010 features the "big four" design criteria described in Chapter Three. The trade-off analysis between and across tiers has apparently been left to the contractors, so it is unclear whether the Army itself has considered a different approach to balancing protection and mobility.

In general, the extra armor that has been added to or is being considered for most existing combat and tactical wheeled vehicles to counter the threats posed in Iraq and Afghanistan has hampered their mobility to some degree, and the additional armor under consideration for all vehicles in development has engendered concerns such as those discussed here. However, it is important to note that protection need not be solely a function of the capabilities of armor or even individual vehicles alone. This broader concept of protection will be discussed later.

---

[4] Andrea Shalal-Esa, "Army, Pentagon at Odds Over New Vehicle Program," Reuters, February 17, 2010.

[5] Headquarters, U.S. Department of the Army, Deputy Chief of Staff for Programs, G-8, *2010 Army Modernization Strategy*, April 23, 2010.

Sensors, Networking, and Power

Another factor that is having an increasing impact on the trade space of current and future combat and tactical wheeled vehicles is the ability to generate electrical power. Specifically, the need to power the plethora of emerging sensors and command, control, and communication (C3) capabilities being designed for the military's fleet of vehicles is putting a significant strain on current vehicle power-generation and -management capabilities. Furthermore, the ability of vehicles to generate power can reduce the requirement for external generators and the trailers that are often used to haul them.

Historically, the HMMWV, MTVR, HEMTT, and similar logistics vehicles have had relatively modest power requirements, primarily for powering military radios. Figure 4.1 depicts the estimated power requirements for a range of tactical and wheeled vehicles based on estimates from 2008.

Figure 4.1
Estimated Power Requirements for Military Vehicles

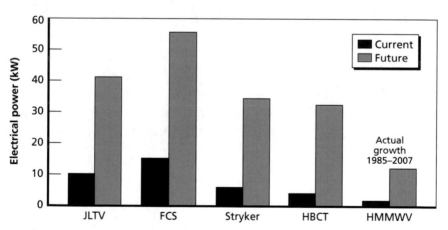

SOURCE: Paul Rogers, "TARDEC's Ground Vehicle Power and Energy Overview," briefing presented at the Michigan Defense and Innovation Symposium, Livonia, Mich., November 17, 2008.
NOTE: HBCT = heavy brigade combat team.
RAND MG1093-4.1

According to the November 2007 CDD for the JLTV (version 2.7a) and the JLTV purchase description available from the TACOM website,[6] the JLTV increment one had a threshold requirement of producing 7 kW of onboard power while the engine is running idle and while moving and an objective of producing 10 kW. The increment 2 had a threshold requirement of 15 kW and an objective of 20 kW. Command and control on the move had an even higher power requirement, with increment 2 producing 20 kW (threshold) and 30 kW (objective), respectively.

The GCV had even more demanding power requirements.[7] Since the cancellation of the GCV RFP, it is not clear what the future power requirement for the GCV will be. However, given the recent trends in power requirements, it is likely to continue to grow.

As shown in Figure 4.1, the demand for more power is affecting the full spectrum of ground combat vehicles and tactical wheeled vehicles. This development reflects the shift away from historical modes of conflict with a "front line" on which most, if not all, force-on-force combat occurred. In modern military operations, logistics and other noncombat vehicles are experiencing the same range of threats as combat vehicles. To respond to the changing threat environment, logistics and tactical wheeled vehicles are adding additional armor as well as enhanced ISR and electronic capabilities (e.g., blue force tracker, jammers, displays, and advanced targeting capabilities) previously reserved for combat vehicles. The demands for additional power have also resulted in increasing demands for fuel and the logistics needed to provide that fuel. Historically, vehicle electronics (vetronics) and electronic systems had a relatively low duty cycle (i.e., the period that the electronics draw power from the vehicle relative to the period that they do not). In other words, the systems were not required to be on or powered for very long periods. For example, the vehicle's electrical system

---

[6] U.S. Army Tank-Automotive and Armaments Command, *Draft Purchase Description (PD) for Joint Light Tactical Vehicle (JLTV) Family of Vehicles*, version 2.3, April 15, 2010. The details are omitted to keep this document available for public distribution. They can be found in U.S. Army, Program Executive Office, Integration, 2010.

[7] Army Capabilities Integration Center, Research, Development, and Engineering Command, and Deputy Chief of Staff G-4, *Power and Energy Strategy White Paper*, April 1, 2010.

(or external generators, when some vehicles were stationary) powered vehicle-mounted military radios. The radios could be monitored for relatively short periods while the engine was off, but the vehicle would have to be started or generators running in order to support the radios for longer periods of time or when the radio was transmitting regularly. However, the duty cycle to support the growing amount of vetronics, sensors, jammers, and communication equipment for most military vehicles is quickly approaching 100 percent. It is becoming necessary to keep vehicle engines running nearly continuously to power the electronic equipment necessary for the vehicle to complete its intended mission.

Several technologies are emerging that are capable of providing additional power. Transmission-integral generators are capable of producing up to 30 kW of power while stationary on an HMMWV platform.[8] Depending on the size of the platform, it may be possible to use this technology to generate significantly larger amounts of power (up to 125 kW).[9]

TARDEC is similarly looking at developments in a range of areas related to ground vehicle power and energy, including power generation, energy storage and batteries, thermal transport and management, and power control and distribution.[10]

Also, given the growth of power requirements for vehicles and the life span of most combat and tactical wheeled vehicles, even if it were possible to meet all of today's power requirements, those systems would be pressed to meet the anticipated new power demands five, ten, or 15 years in the future. Therefore, to be prepared for what is likely to be an ever-growing series of systems, it will be necessary to develop vehicles with an open architecture capable of easily integrating new

---

[8] DRS Technologies, "HMMWV Power Now, Anywhere!" Huntsville, Ala., August 19, 2010b. These generators are integrated into the vehicle transmission.

[9] "DRS and Allison Transmission Announce Strategic Partnership," ASD News, November 2, 2010.

[10] U.S. Army Tank Automotive Research, Development, and Engineering Center, "Overview: Tank Automotive Research, Development & Engineering Center," briefing presented at the Automotive Supplier Technology Forum, Troy, Mich., June 22, 2010b.

systems as they become available or necessary. It will also help enable modularity to support the swapping of the "plug-and-play" vetronics, sensors, and C3 systems needed to support the mission.

An additional benefit of onboard power generation is that it does away with the need for many generators (and the trailers that often carry them). This removes a significant logistical burden from the force and decreases the space needed in strategic and operational transport systems.

Up until this point, we have primarily stressed the increasing demands for power to support the growing number of sensors, C3 systems, and vetronics systems that are being integrated into the current and planned fleet of combat and tactical wheeled vehicles. Another important design consideration is the added complexity and challenges associated with integrating them into the vehicle. As mentioned earlier, most previous vehicles may have had a single radio, which not only had minimal power requirements but also was relatively easy to install and integrate into the platform. Today's current and planned vehicle fleet must support a wide variety of information systems (e.g., voice and data radios, battle command systems, counter-IED devices, combat ID systems, Global Positioning System devices) and operate with minimal interference. Despite vetronics and data buses, these requirements are stressing not only the technical limits of power generation to enable them but also the ability of system integrators and vehicle designers to integrate the systems into vehicle platforms.

Engine Design and Fuel Efficiency

Another trade-off among vehicle design criteria has to do with engine design and fuel efficiency. In the Marine Corps, the MTVR uses 50 percent of all the fuel consumed on the battlefield. News reports indicate that the fully burdened cost of fuel (FBCF) in some conflict areas can be as high at $400 per gallon for JP8, which is used for both military aircraft and ground vehicles.[11] However, some estimates range from as low as $20 per gallon up into the thousands.[12] The extremely high prices are a result of the challenges and resources necessary for transporting fuel to difficult, dangerous, and often remote locations, such as parts of Afghanistan. Methods for estimating the FBCF include "the total life-cycle cost of all people and assets required to move and protect fuel from the point of sale to the end user."[13]

There are several approaches for reducing the FBCF. One is to develop more efficient engines, such as integrated hybrid-electric technologies or other engine designs. Another potential and perhaps complementary approach would involve altering engine designs to accommodate local fuels found in theater. These steps would significantly reduce the need to transport fuel from other locations to the area of operations.

In this regard, one trade-off has to do with the availability of technologies capable of improving the engine efficiency of military vehicles (e.g., FCS was developing a hybrid engine for its manned ground vehicles). Many of these technologies are still in the development phase and must mature further before they are capable of being integrated into military platforms like the MTVR. In 2008, Chris DiPetto from the Office of the Deputy Under Secretary of Defense for Acquisition and Technology spoke before the House Committee on Armed Services, Readiness Subcommittee, on the energy risks and challenges faced by DoD. Specifically,

> when we design our future capabilities in the Pentagon or at the major Service materiel commands and elsewhere, logistics demand of our capability choices are not addressed until after we

---

[11] Roxana Tiron, "$400 per Gallon Gas to Drive Debate Over Cost of War in Afghanistan," *The Hill*, October 15, 2009.

[12] C. Todd Lopez, "Saving Energy Saves Lives Says New Army Exec," Army News Service, October 13, 2010c. However, the Logistics Integration Agency's most recent estimates are $20–$25 per gallon, according to Army sources.

[13] Chris DiPetto, Office of the Deputy Under Secretary of Defense for Acquisition and Technology, "DoD Energy Demand: Addressing the Unintended Consequences," September 2008b.

have decided on what that performance our platforms or combat units should have [sic]. Stated more simply, our force planning processes almost always plug fuel logistics in at the back end, after the capability we want is designed. The result is that we plan capabilities and systems ignorant to the combat support "tail" we are creating. . . . Further, we make decisions on the un-refueled range and payload and loiter time of platform types, but at no point [do] the force development processes consider whether it's worth it to reduce the logistics demand to gain unit or theater deployability, vulnerability or sustainability benefits. Finally, we have little to no analysis on which to determine what it's worth to the larger force to invest in fuel efficiency technologies. We're largely allocating investment based on military or technical experience, not on modeling, wargaming, trend analysis or other accepted tools.[14]

This statement very succinctly makes the point that there are important implications across several "stovepipes" that are not taken into consideration by the current acquisition processes. This fact, in turn, affects the overall costs that DoD must bear and the funding that Congress must provide.

Cost and Schedule

Historically, program managers measured and evaluated the execution of a program across three general parameters: cost, schedule, and performance. Changes along one parameter often had consequences for the other two. Typically, if a program did not execute as expected, and if the capabilities associated with the program were deemed critical, modifications to the program's schedule or costs (sometimes severe modifications) were incurred to achieve the desired performance.

One example of this is the EFV. The original mission needs statement for a new platform to replace the Marine Corps amphibious

---

[14] Chris DiPetto, Office of the Deputy Under Secretary of Defense for Acquisition and Technology, testimony before the House Committee on Armed Services, Subcommittee on Readiness, March 13, 2008a.

assault vehicle was written in 1990. The contract was awarded to General Dynamics in 1998. Since entering the SDD phase in December 2000, the program has encountered several technical challenges, and as a result, it has incurred significant cost and schedule growth as it seeks to achieve the performance reflected in the vehicle's specifications. As noted in Chapter Three, the original unit cost per vehicle was estimated to be approximately $8.5 million, while current production cost per EFV is estimated to be $16 million, with a total acquisition cost of about $22 million per EFV. Similarly, when the EFV program entered the SDD phase, the original plan was to complete SDD by October 2003. Subsequently, the second round of EFV prototypes was delivered to the Marine Corps for testing in September 2010. The current plan, assuming all goes as predicted, is for the EFV to enter limited production by 2012 and for the program to field 573 vehicles by 2026.[15] While the EFV is one of the most notable examples of cost growth and schedule delay for a ground vehicle, it is hardly the only program in this situation.

Even programs still within the technology development phase have the potential to incur added costs or delays. For instance, the original cost estimate for the JLTV, to replace every HMMWV in the fleet, was approximately $250,000 per vehicle. However, the current estimate for the JLTV is more than $300,000 per vehicle,[16] and some estimates run as high as $418,000.[17] This is significantly higher than the recap costs for an unarmored HMMWV, which is estimated to be about $55,000 per vehicle, or for the up-armored HMMWV at about $105,000–$130,000 per vehicle.[18] A recent interview with GEN Peter Chiarelli did suggest that there were some improvements in cost

---

[15] Masato Itoh, "New EFV Prototype Tests at Camp Pendleton," *Ground Combat Technology*, Vol. 1, No. 2, August 2010.

[16] Kate Brannen, "U.S. Army Reaffirms JLTV Commitment," *DefenseNews*, October 26, 2010f.

[17] Feickert, 2010.

[18] C. Todd Lopez, "M-ATVs to Replace Humvees in Afghanistan, Vice Says," Army News Service, April 15, 2010a.

with the JLTV program,[19] perhaps getting closer to the original target of $250,000 per vehicle.[20]

Recent policy documents and memoranda from the Under Secretary of Defense for Acquisition, Technology, and Logistics have made it clear that, in order to address the escalating costs and prolonged schedules for major acquisition programs, program managers and program executive officers should "do more without more."[21] In his memo, Under Secretary Carter cites the GCV as an example of the importance of addressing schedule directly. He mentions that the initial acquisition plan had assumed ten years to complete first production. According to Carter's memo, DoD determined that the GCV program should have a seven-year schedule to the first production plan. He also stated that, for the GCV and, by extension, all DoD programs, "requirements and technology level for the first block of GCVs will have to fit this schedule, not the other way around."[22]

Robotics and Other Emerging Technologies

The trade-offs described here consider current engineering solutions and developing technologies to achieve some optimal mix of vehicle design criteria. However, over the coming years, new technologies may emerge that have the potential to radically alter the balance among these various trade-offs. One such potential technology is robotics and the use of autonomous systems (e.g., unmanned aerial and ground vehicles). The ability to remove the crew from a vehicle has vast implications for vehicle designs and, potentially, for new missions. In the same way that unmanned systems have transformed how the Air Force conducts ISR missions and how the Army detects IEDs, unmanned, autonomous ground systems have the potential to redesign and reinvent ground

---

[19] Kate Brannen, "Gen. Peter Chiarelli: Vice Chief of Staff, U.S. Army," *Defense News*, October 25, 2010e.

[20] Feickert, 2010, p. 5.

[21] Carter, 2010c.

[22] Carter, 2010c.

vehicles. For example, autonomous following technology demonstrations, in which a single, manned vehicle leads a convoy of unmanned vehicles, could significantly reduce protection requirements, increase vehicle payload, and minimize the need for onboard communication and situational awareness equipment for the unmanned vehicles in the convoy.

Similarly, the ability for current or new military vehicles to interface and interact with autonomous systems may provide new options for the design of vehicles—potentially allowing specific sensor packages or communication systems to be offloaded to loitering unmanned aerial vehicles or other companion vehicles.

Unmanned systems would also radically alter the trade space for vehicles. By eliminating the need to protect soldiers and marines in some vehicles, the weight requirement would lessen, with the attendant gains in all other variables that trade off with weight. Similarly, fuel consumption, other logistics requirements, and strategic mobility could be significantly enhanced.

It is difficult to predict exactly how robotics or other emerging technologies may ultimately transform the design and development of military vehicles. The task is more difficult if one considers that ground vehicles are only one type of system in a expansive system of systems that includes other ground vehicles, dismounted soldiers, and (potentially) air and space assets, as well as systems located thousands of miles away from the area of operation. What is certain is that these emerging technologies *will* shift the balance of trade-offs, and *will* ultimately affect the design of the future combat and tactical wheeled vehicle fleet. The timing of this shift is an issue of relevance to Congress as it considers the future vehicle fleet.

Implications

As the dyads of trade-offs discussed in this chapter suggest, trade-offs become more difficult as vehicle designs seek to include more functionality. Although we did not note it explicitly, mobility generally serves as a bill-payer for virtually everything else in a vehicle's design, because virtually every other aspect of the design, when enhanced, adds weight or requires power, which ultimately detracts from mobility. Increasingly difficult trades within a vehicle's design imply schedule delays and cost growth.

Moreover, the trade space on a vehicle is zero-sum in the sense that enhancements in one design criterion lead, necessarily, to constraints on others. When minimum essential performance across all design criteria cannot be achieved on the vehicle in question, then engineers must look to off-vehicle solutions to make up the difference. Doing so means that they must consider the full range of DOTMLPF options and the ability of other vehicles, perhaps including unmanned vehicles, to provide the necessary capabilities.

Modeling and simulation can be helpful in this regard by identifying the limits to on-vehicle trade-offs and representing possible off-vehicle effects. The next chapter considers how M&S might be enhanced to improve support to vehicle development.

CHAPTER FIVE

Modeling and Simulation for Ground Vehicle Modernization and Capability Development

Modeling and simulation have evolved to play an increasingly important role in the identification of operational and program requirements, as well as the development, refinement, production, and fielding of U.S. military ground vehicles.[1] Each successive generation of platforms has seen greater involvement of computational analysis, detailed modeling, force-on-force simulation, and field experimentation. This is partly due to the rapidly rising costs, risks, and time and manpower requirements of physical field tests and prototyping and partly because of the increasing capabilities of M&S in terms of accuracy, scope, and realism.

The principal focus of this chapter is M&S in the context of capability development and evaluation, a process that is used before, during, and after technology development (TD) and EMD.[2] More formally, TD and EMD span the time between milestone A and milestone C in DoD life-cycle terminology. TD and EMD require a different set of modeling and simulation tools to manage the technical aspects, cost, schedule, and risk of programs in these phases. For completeness, we briefly cover these different M&S tools and their use in a later section, "The Use of Models and Simulation in the Management of Technol-

---

[1] Throughout this monograph, we use the term *modeling and simulation*, or M&S, as shorthand for a wide range of analytic tools, spanning such functions as computational analysis, engineering models, SME interactions, constructive and virtual simulations, and field tests.

[2] Several of the programs discussed in this monograph took place before the term *EMD* was coined. EMD is a refined version of system design and development (SDD), which was used for those earlier programs.

ogy Development and Engineering and Manufacturing Development Programs." Prior to and after that section, this chapter addresses M&S in the context of capability development and evaluation.

In the discussion that follows, we distinguish between operational M&S and programmatic M&S for capability development, as illustrated in Figure 5.1. With respect to operational M&S, it is useful at the start to briefly define key elements of these efforts and make important observations. First, the purpose of operational M&S is to help DoD and armed services concept developers understand how conflicts may play out in the future and what capabilities are needed for success. To do this, the M&S community must have models that adequately depict expected conflict scenarios and how forces and systems will interact to prosecute those conflicts. In particular, no one system will operate alone, and so the ability of M&S to assess an individual system—be it a main battle tank, a heavy truck, or a unmanned aerial system—in isolation is limited. For example, vehicle protection from IEDs is accomplished through the combined contributions of sensors that look for IEDs, various capabilities that neutralize IEDs, and the ability of vehicles to defend their crews from an IED blast. Similarly, protection from direct fire systems depends on a number of systems that sense, defend, and protect. Only a consideration of the context of how a conflict is to be fought by forces and multiple systems on a battlefield can yield insights into the operational requirements for a system.

Programmatic M&S focuses on the tools that help define program requirements and evaluate program performance. The primary role of programmatic M&S for ground vehicle modernization is to refine and evaluate the various requirements and capabilities that were introduced in earlier chapters, with particular emphasis on complex trade-off analyses, system-of-system explorations, identification of key technology elements, and estimation of integration risks.[3]

---

[3]   Of special importance will be how M&S is called on to assess the complex interactions between systems and to manage the rapid interchange of modular upgrades in equipment subsystems, such as radios, displays, power distribution equipment, sensors, and defensive suites. These plug-and-play subsystems are exhibiting shorter and shorter life cycles and greater interconnectedness.

Figure 5.1
M&S Supports System Development and Fielding

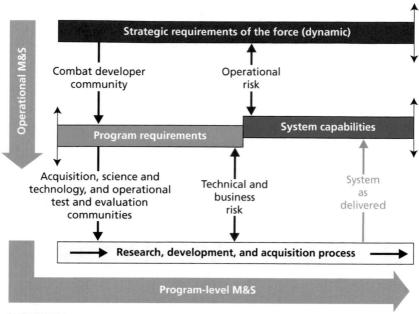

RAND *MG1093-5.1*

By looking at operational and programmatic M&S, we are able to explore how modeling, simulation, and other forms of analysis can help answer the question, "Will the future DoD combat and tactical wheeled fleet be adequate for U.S. strategic requirements?" We also consider shortcomings in current M&S capabilities and recommend improvements for the near and far terms.

The discussions and recommendations offered here are targeted toward two groups: (1) the analysis and testing community, to help it evolve the tools and procedures to keep step with a rapidly changing battlefield, and (2) acquisition decisionmakers, who must understand the M&S process and utilize the findings.

The scope of this research dictates that most of our focus be on programmatic M&S. However, there will points in the discussion that follow in which the ability to integrate M&S tools in support of decisionmakers will be important, so we will have occasion to reference both at times.

The Challenge

Current efforts to keep up with rapidly changing requirements on the battlefield are struggling. The combination of an adaptive enemy, rapidly changing conditions and goals, and the introduction of nontraditional acquisition channels necessitates a responsive process. Use of more capable M&S tools should increase speed and responsiveness in several different ways. In particular, it could

- shorten the time between identification of a requirement and delivery of a solution
- allow new capabilities to be fielded quickly, leveraging breakthroughs in DoD's and the services' science and technology programs and technology-based initiatives
- enable the continuous modernization of equipment, using procurement, recapitalization, and divestment
- help with the implementation of comprehensive, affordable portfolio strategies for the vehicle fleet.[4]

The need for flexibility and responsiveness in development and acquisition is also seen in the many different types of questions now being faced by ground vehicle developers, including those related to issues of robustness to new enemy countermeasures; changing interoperability requirements and other conflicts between systems; trade-offs among protection, agility, and situational awareness; and changes in tactics with shifts in missions and rules of engagement. These questions relate to all aspects of the vehicles (along with their operations, tactics, and missions). The relative importance of each issue depends on the phase of development.

---

[4] These points are paraphrased from the *2010 Army Modernization Strategy* (Headquarters, U.S. Department of the Army, Deputy Chief of Staff for Programs, G-8, 2010).

What M&S Tools Are Available to the Analyst?

Before discussing the specifics of these challenges, it is useful to look at which tools and processes are available to the analytical community today. Figure 5.2 presents some of the major classes of tools and procedures available to developers and decisionmakers. At the top of the figure, engineering models represent such physical subsystems as suspension, power, communications, and armor. These detailed models characterize a specific system and show how it performs against a threat or under a specific condition. They are usually very localized, often modeling the system only during a snapshot of action.

Constructive force-on-force simulations model a group of vehicles and other entities and calculate outcomes (e.g., movements, detections, shots, kills) over the course of a mission segment or scenario. The visual representation is often an abstracted set of icons on a map display, and the action is often scripted and noninteractive. This type of simulation is often the workhorse for evaluation of options, as it tends to be veri-

Figure 5.2
General Classes of Analytic Tools

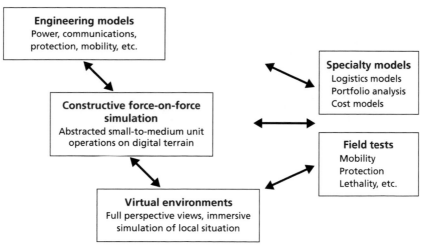

fied, validated, and accredited (VVAed) through intensive testing of the software and outputs.[5]

Virtual environments have a more immersive quality than constructive simulations, allowing the human user to interact with the environment. Scenes are typically shown in perspective, 3D view, and interactions may involve realistic interfaces, such as steering wheels and pedals. These tools have often been adapted from the high-resolution commercial gaming world but have seldom been extensively VVAed.

Specialty models, such as logistics models, may use spreadsheet calculations or other computational procedures to estimate fuel use, refueling intervals, and payload delivered per unit time. Some specialty models may also attempt to optimize parameters with respect to costs or burdens. An example of such an optimization procedure is the Portfolio Analysis Tool, used for such purposes as missile defense and acquisition decisionmaking.[6]

Field tests are often considered the gold standard for analysis. These are typically physical reconstructions of the actual mission environment, using training sites or test areas to see how systems perform in challenging environments. They also contribute significantly to the body of information available to support model VVAs.

These many processes may also be augmented with sand table exercises or map-based campaigns, with SMEs working through hypothetical situations. Many of these tools may be used, in turn, to provide inputs to the SME exercises.

The many tools have very different burdens of operation and levels of validation and effectiveness. Table 5.1 summarizes these characteristics with respect to capability development in an initial, rough manner,

---

[5]   *VVA*, as used in this document, refers to models and simulations. The process ensures that the model does what it is supposed to do and does it accurately. This involves both internal and external tests of the consistency, functionality, and credibility of software, components, and outputs.

[6]   Paul Dreyer and Paul K. Davis, *A Portfolio-Analysis Tool for Missile Defense: Methodology and User's Manual*, Santa Monica, Calif.: RAND Corporation, TR-262-MDA, 2005, and by the same authors, *RAND's Portfolio Analysis Tool (PAT): Theory, Methods, and Reference Manual*, Santa Monica, Calif.: RAND Corporation, TR-756-OSD, 2009.

Table 5.1
A Rough Characterization of the Types of M&S Tools

Tool	Development Time	Staffing	Iteration Speed	Realism	VVA
Engineering models	Days to weeks	Low	Fast	Moderate	Good
Constructive simulation	Weeks to months	Moderate	Fast	Moderate	Good
Virtual environments	Months	Extensive	Moderate	High	Very limited
Specialty models	Days to weeks	Low-moderate	Fast	Low	Limited
Field tests and experiments	Months	Extensive	Slow	High	Good

using our estimates and those of experts who participated in the study workshop.

TRADOC researchers further elaborated on the cost burden of many of the different venues. They assumed that the simplest seminar war game, walking through a single sequence of events with no aids or special tools, could be used as a baseline cost of 1.0. Staffing experiments and war games examining alternatives using maps and other physical aids were estimated to cost between 1.25 and 1.75 times the baseline cost. Computer-assisted mapping experiments, which we interpreted as equivalent to using constructive simulation to examine multiple options, were estimated to be four times as expensive to conduct as the baseline. Human-in-the-loop experiments, finally, could be thought of as the use of virtual environment analysis. This latter approach would be very time-consuming, personnel-intensive, and expensive, typically 25 times the baseline cost.[7]

When considering M&S for analysis, the importance of SMEs cannot be overstated. The tools are an abstraction of military operations and in many cases cannot accurately portray or quantify key phenomena. This is particularly true of psychological aspects of combat,

_____

[7] U.S. Army Training and Doctrine Command Regulation 71-120, Concept Development, Experimentation, and Requirements Determination, October 6, 2009.

such as motivation, stress, and fatigue. Often, battle-stopping conditions are somewhat arbitrarily set to values such as loss of 40 percent of the force, with the assumption that the unit is combat-ineffective below this number. In fact, many units may still be operational while others may be combat-ineffective at a point well before this threshold. SMEs are needed to estimate these events and can be very effective in seeing turning points in engagements. They can also be very useful for quickly screening options and providing insights in terms of the credibility of analyses. In fact, the use of SMEs provides some inherent accreditation, depending on the group's experience and capability.

M&S and the Development Cycle

Each of the different stages of system development and acquisition has a different set of needs and constraints for M&S, with most of the process acting in a form of "necking down" from very broad, exploratory analysis to much more directed and specific computation. The process is not monotonic in nature, but often moves in an iterative manner, with many iterations between what would otherwise be sequential steps. As a result, M&S support must be flexible in its capability and application. This is especially true for the new starts, such as JLTV, EFV, and GCV. These new starts follow the full sequence of development and production stages. With time, systems become more defined in their capabilities, specifications, and missions, and the associated M&S becomes more focused and specific. Recapitalizations, in fact, may involve only limited explorations of engineering options and interoperability concerns.

Concept/Material Development Decision

Here, the objectives of M&S are to aid with the functional area analysis to arrive at a variety of potential options and to establish rough measures of effectiveness and performance. The modeling approach is then to quickly screen and examine many different concepts and tactics, determining whether critical constraints or performance characteristics have not been met. The most efficient tools in this phase are often SME exercises, simple computational estimates, and other forms of qualitative screening. Formal simulations and high-fidelity models are often an unnecessary complication at this point. As the material development decision nears, the milestone decision authority determines the scope of the analysis of alternatives (AoA).[8] In the AoA, KPPs are established for each alternative, and more systematic analyses are performed using high-resolution simulation and engineering models.

Technology Development Decision

Throughout the TD phase, further M&S is performed to refine the concepts. In this and succeeding phases, either the AoA is updated or a new one is conducted.[9] As the requirements become more specific, M&S makes greater use of engineering models (examining and representing tracks, sensors, communication models, and other systems) as inputs to the force-on-force models. The favored tools here also expand to include virtual environments and specialty models, to more carefully refine the system and its associated tactics, techniques, and procedures (TTPs). At this point, only a small number of specific capability options remain for evaluation, and the scenarios should shift to more stressing and definitive ones.

The Use of M&S in the Management of Technology Development and Engineering and Manufacturing Development Programs

In this section, we revisit the TD and EMD phases of a program's life cycle and address M&S as used to manage the cost, schedule, risk, and technical aspects of major (ACAT I-D) programs.

At the start of EMD (before milestone B), an RFP is issued to solicit proposals for a system that meets the department's requirements. These typically include (1) financial requirements in the form of budget available for EMD and production and the unit production, operating, training, and support costs; (2) schedule requirements, including key

---

[8] Chairman of the Joint Chiefs of Staff Instruction 3170.01G, Joint Capabilities Integration and Development System, March 1, 2009.

[9] Marc Greenberg and James Gates, "Analysis of Alternatives," Washington, D.C.: National Defense University, April 2006.

delivery dates; (3) the identification of potential high-risk areas and a plan to mitigate them; and (4) a system specification that identifies the technical performance and means to verify technical performance of the systems that the bidder will deliver as engineering prototypes in EMD, early production articles in low-rate initial production, and full-production articles in full-rate production.

The degree of detail in the system specification varies with the product sought. If the product is similar to an existing product, the system specification can be highly detailed. If DoD wants to evaluate bidders proposing different solutions, the system specification will be more sparse, allowing more freedom and variety in the proposals that are submitted, but at a minimum, it will still include required KPPs, key system attributes, and other specifics to bound the proposals.

The bidders add significant detail in their proposals in all these areas as they form the blueprint for what the successful bidder will provide under contract. The more specific the contract, the better the position of the contractor in knowing what to build, what engineering and manufacturing processes to employ, and the degree of confidence in the bid cost, schedule, and performance. Negotiations between bidders and the DoD department or command that issued the RFP add more detail, increasing the government's knowledge of the bid and thereby reducing the likelihood of selecting the wrong proposal.

DoD invokes a milestone B decision process when it believes it has identified a best bid that is suitably low-risk, affordable, and timely and that provides the capabilities it wants. If endorsed by the milestone B process, an EMD contract is awarded.

The end state of an EMD contract is not perfectly known at the beginning. If it were, production could start at that point. During the next five or more years, product design detail will increase as lower-tier design trade-offs and decisions are made and engineering and manufacturing problems are encountered and solved.

The contract contains a description of the M&S tools that will be used to assess and control cost, schedule, risk, and technical performance. These tools will become increasingly specialized and mature with time as product detail emerges through design, test, and evaluation. In discussing these topics, it must be remembered that cost, schedule, risk, and technical performance can be highly coupled and interactive. A technical performance issue, such as a system integration problem, can affect cost, schedule, and risk; a schedule problem, such as a late delivery with some deficiencies, can affect costs, schedule, and technical performance; a risk that cannot be adequately mitigated can affect cost, schedule, and technical performance; and an external program budget change due to new congressional, DoD, or departmental priorities will most likely affect many if not all of these program parameters.

The contract, further, contains an integrated master program plan, a model of how the program will proceed. It also contains an integrated master schedule that models when program events will occur. It contains a work breakdown schedule that models the program organization and how the work, responsibilities, authorities, accountabilities, and budgets are organizationally distributed, defining the contributions to scheduled events that the organizational components need to satisfy. This is a model of the organization and the costs and work products expected from its components, and how these work products are used. The contract contains an earned value management system (EVMS) that was constructed using the integrated master plan, integrated master schedule, and work breakdown schedule and links them. The EVMS is a mega-model that assesses program progress against its planned schedule, budget, and technical accomplishments, and it flags issues so that they can be addressed.[10] The risk management plan, a model that tracks risks, efforts to mitigate them, and the consequences of those efforts, is linked with a part of EVMS so that serious risks are proactively managed—both known risks and those that emerge as the program proceeds.

These models and their use become more complex with time as program detail is developed, but also for another reason: The system being developed has many layers, or tiers (see Figure 5.3), some of which are the domain of subcontractors, component developers, and other suppliers. To the extent practical, the subcontractors and suppliers are

---

[10] Under Secretary Carter recently stated that he considers EVMS to be the most reliable indicator of program health.

required to maintain cost, schedule, risk, and performance models that are compatible with those of the prime contractor so that their program data can be integrated into the EVMS mega-model, because their program performance affects that of the prime contractor. The net result is that computer models that can continuously integrate the effects of all these interactive data elements from multiple sources and display the consequences as required for the operation of EVMS and other complex management tools have become a business enterprise service with several competing vendors offering access—not ownership—to their computing environments and programs for this purpose. These enterprise-level tools are computer-intensive dynamic models, driven by current status data, that provide a simulated representation of the state of the EMD program for management purposes.

The tiered nature of complex engineering and manufacturing development is illustrated in Figure 5.3.[11] Typically, there are seven or more tiers that need to be planned and managed in the face of design and risk uncertainty, but, for simplicity, only three are shown. They are components that make up subsystems that make up systems. Time proceeds from the top of the left branch of the V, beginning with the system tier to the subsystem tier to the component tier to the make-or-buy process (the design path) and then up to component to subsystem and to system (the manufacture and integration path).

At the beginning of an EMD program, the system—such as an Army tactical wheeled vehicle—has been proposed and defined well enough that its performance can be predicted with some accuracy based on a constructive simulation, generally called the *system simulation*. The system simulation includes physics-based simulations of the environment, such as how a projectile fired from a turret-mounted gun will fly toward a target or the signature provided by enemy vehicles in applicable parts of the electromagnetic spectrum that can be sensed by the vehicle's sensors. It also includes simulations of the technolo-

---

[11] The V paradigm shown in the figure is a simplified version of the activities that take place in an ACAT I-D program. Only those activities associated with delivering the prime product (say, the tactical wheeled vehicle, in the context of this monograph) are represented as the system of interest. But in reality, an EMD program would include its support system, training system, and manufacturing implementation.

Figure 5.3
The "V" Paradigm

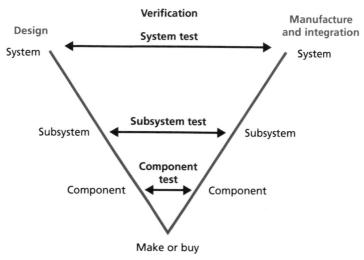

gies, subsystems, and the system itself and external factors, such as the networked information grid with which the system interacts, so that it becomes the means by which the system performance can be estimated and compared to that required by the system specification. As time proceeds and progresses down the left side of the V paradigm, the system design and trade-off process produces more-detailed, better-informed feedback at the system, subsystem, and component levels, which is fed into the system simulation to improve its capability to represent the system. When the system is sufficiently defined, with every subsystem and component well-enough understood and having its own specification, then make-or-buy decisions are made—that is, the prime contractor decides which subsystems and components to make and which to buy.[12] Typically, some of the subsystems and most of the components are bought. The subsystem providers of the bought subsys-

---

[12] This does not happen at one point in time, but over an extended period, depending on the maturity of the subsystems and components. And generally, certain subsystem and component suppliers that are part of the proposed system are selected and presented as such in the bidder's EMD proposal and then become part of the contracted system.

tems are responsible for its components, and the prime, as the maker of the system, is responsible for integrating the system and for the performance of the system as a whole.

Following the make-or-buy decision, prototype components and, later, prototype subsystems become available and are tested against their specifications to ensure that they perform as required, with adjustments made as required.[13] Time-wise, the program is moving up the right side of the V paradigm.

The system simulation converts from a constructive simulation to a mixed hardware/software simulation, where *hardware* refers to the replacement of computer code in the constructive simulation by the prototype hardware (actually, hardware and software) that it was meant to represent, and *software* refers to those portions of the constructive simulation that remain in place. The host for the constructive simulation is a computer or system of computers. The host for the mixed hardware/software simulation is the system integration laboratory (SIL), which contains the hardware and software elements of the system simulation. The SIL also contains emulators with which some of the hardware elements interact, such as an infrared target simulator that characterizes the infrared sensors represented in hardware in the SIL. The SIL can also include consoles and operators so that the human-machine interface resident in the system is present.

As the SIL (and other program activities) uncovers problems, design changes are introduced to correct them, and thus the V paradigm becomes more realistic as prototype subsystems, for example, that have been designed, manufactured, and integrated, return to an earlier state of design, integration, and testing. This is inherent in the iterative nature of design. But the SIL continues to grow in its fidelity to represent the performance of the system until it is finally superseded by the prototype systems that undergo engineering testing as part of EMD. But even then, the life and utility of the SIL continues. Engineering testing is conducted with prototype systems on instrumented test ranges and is very expensive. For this reason, some aspects

---

[13] In EMD, the deliverables are used to construct system prototypes that are tested in the engineering test phase at the end of EMD.

of system performance are not tested. In these instances, system testing is used to calibrate the SIL, which is then used to evaluate those untested aspects and other excursions undertaken, for example, to evaluate the system modifications necessary to address problems that arise after system testing or opportunities to improve system performance for system performance growth.

Up to this point, this section has emphasized the use of models and simulation for the management of program development in EMD programs. But EMD also develops the means of manufacture of the systems designed and prototyped in EMD. Even in the most compact form, this subject is as complicated as what we have discussed so far and beyond the space we can devote to it in this document. We can only suggest that the V paradigm applies, that manufacturing is a system, and that its subsystems include the manufacturing machines themselves, their physical and functional relationship to one another, the processes they employ, the process flows they can support, and the processes that they work with and depend on at the levels of supply chain, inventory, and work in progress, and the training of manufacturing personnel. M&S is employed extensively to optimize factory process flow, minimize downtime, evaluate lean opportunities, and manage the manufacturing process.

Similarly, we have not discussed the TD process that precedes EMD and provides the mature technologies that are needed at the initiation of EMD, but here, the analogies with the V paradigm are easier to make than with manufacturing. Technologies can be small essential things, like focal plane arrays. In this case, they are the system shown in the V paradigm, and their subsystems and components can be derived from how the technologies are constructed. More generally, technologies are embedded in what become the subsystems of ACAT I-D and other program-of-record systems, and it is these that become the systems of TD programs. These subsystems are developed using the same system management and development processes discussed for EMD in this section, but with less rigor in keeping with the more specialized requirements, lesser complexity, and smaller budgets for TD. In the context of tactical wheeled vehicles, a technology development program might consist of one of its sensors, a subsystem of the tactical wheeled vehicle program, or the sensing head (the sensor less signal processor, and the mechanical pointing and tracking system with its gimbals and actuators), a major subsystem of a subsystem of a tactical wheeled vehicle program. However done, the objective of such a TD program is suitable technical maturity, typically TRL 6 or 7, as is required prior to the start of EMD.[14]

The Recapitalization Process

An existing system may be modified to meet changed operational conditions or reduce costs. Often, these changes are large enough to trigger new requirements, which may require specification of new measures of effectiveness and measures of performance. The platform is sometimes many decades old, such as the M109, and may suffer from inadequate acceleration, off-road capability, reliability, networking, or payload. Such a comprehensive redesign may require restarting almost the entire 5000-series development process, but simpler changes may enter the process much later, possibly even as changes in production. M&S will often be used only to screen and examine new component technologies, tactics, and integration options, with the goal of determining whether there is sufficient performance improvement to justify the cost of the recap. The favored M&S tools in this phase are typically subsystem and systems engineering models, the SIL, and system simulation, along with specific scenarios for the modifications. Field tests may suffice for demonstration of the desired performance.

Nonstandard Acquisition

Many new rapid acquisition developments have been introduced in the past few years, including those arising from the Rapid Equipping Force, the Rapid Fielding Initiative, the Joint Urgent Operational Needs Statement, and material solutions developed by the Asymmetric Warfare Group. The Capabilities Development Rapid Transition process is designed to move these initiatives and nonstandard equipment from operational theaters into long-term capabilities and programs

---

[14] This chapter now returns to discussing M&S for capability development and evaluation.

of record.[15] This means that rapid acquisition developments, many of which were moved into the field and achieved some degree of success, could, with the aid of M&S, transition into the deliberate development and acquisition process. This has already been done in some cases. The form of M&S needed would depend on the phase of entry and the degree of interaction and interoperability with other existing and planned systems.

A special case arises because of the increasing adoption of modular, common components. Examples of this type of component are common radios, high-voltage electrical harnesses, active protection systems (APSs), remote weapon systems, and thermal sensor packages. Many of these modular systems can be added to virtually any tactical wheeled or ground combat vehicle in the fleet. They can apply to restarts, nonstandard acquisitions, and even new starts. M&S procedures for evaluating these components should be done for several applications at once and assume sufficiently general conditions to allow their evaluation to apply to many different platforms.

The Need for Rapid and Appropriate Scenario Generation and Validation for Capability Development

Future ground combat and tactical wheeled vehicles will have to carry out missions in conflicts ranging from humanitarian operations to major regional conflicts involving peer competitors. The M&S scenarios thus need to exercise the full range of conditions and focus on those that are the most stressing in order to highlight key requirements and gaps.

The five characteristics of combat scenarios, shown in Figure 5.4, might be described as intensity, scale, exposure, terrain, and weather. All need to be reflected to some degree in the choice of scenarios and vignettes. In general, the more stressing levels involve variants of combat operations in degraded weather and rough terrain. It should

---

[15] See U.S. Army Training and Doctrine Command, "Capabilities Development for Rapid Transition (CDRT)," in *2010 Army Posture Statement*, December 2009.

Figure 5.4
Some Key Characteristics of Combat Scenarios, Exercising the Range of Operations for Military Ground Vehicles

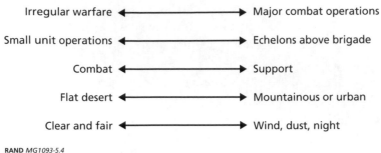

be apparent that a given system must be tested against a wide range of such conditions. Unfortunately, this has not always been the case.

TRADOC and its Analysis Center (TRAC) are the Army's primary originators and vetting groups for new scenarios. TRADOC develops concepts, CDDs, and CPDs and directs AoAs, while TRAC and the Army Capabilities Integration Center in TRADOC are responsible for developing standard scenarios and overseeing M&S, producing concepts, and leading experimentation and requirements specification.[16] TRADOC and TRAC work in coordination with many other commands and agencies to identify, analyze, test, and develop new warfighting systems and concepts.[17] Not surprisingly, this process is not

---

[16] U.S. Army Training and Doctrine Command Regulation 71-120. See also Headquarters, U.S. Department of the Army, *Simulation Support Planning and Plans*, Pamphlet 5-12, March 2, 2005.

[17] Headquarters, U.S. Department of the Army (G-3/5/7, G-8, and G-9), the Army M&S Office, and RAND Arroyo Center are all responsible for coordinating operations, plans, and testing and for participating in war games and modeling. Headquarters, U.S. Department of the Army, and the Office of the Secretary of Defense produce defense planning scenarios. The U.S. Army Research, Development, and Engineering Command (RDECOM); the Defense Advanced Research Projects Agency; the Army's Engineer Research and Development Center; the Office of the Secretary of the Army for Acquisition, Logistics, and Technology; the Center for Army Analysis; and the U.S. Army Test and Evaluation Command are part of the community of practice. Finally, the former U.S. Joint Forces Command coordinated joint experimentation support and conducted exercises (this function has been moved to other joint commands).

always smooth, and many of the agencies use very different scenarios, assumptions, and procedures in their parts of the process.

An examination of several upcoming plans for AoAs provide an idea of the types of scenarios proposed by TRAC. The draft GCV and JLTV AoA plans each have roughly a half-dozen scenarios involving Operation Iraqi Freedom, Operation Enduring Freedom, Northeast Asia, Southwest Asia, and U.S. Africa Command terrain and threats. They involve attacks and defenses from the small-unit to brigade combat team level.

Looking at the set of descriptors listed earlier for a full range of scenarios, we find that there appear to be several missing aspects. Representative urban operations should be added, ranging from protection of the population or humanitarian efforts to clearing of buildings. Protection of bases, such as forward operating bases, combat outposts, fire bases, and observation posts, are important missions but are not present in the draft plans. Lessons learned from such incidents as Combat Outpost Keating and Vehicle Patrol Base Wanat showed the need for improved vehicle protection, networking, lethality, and mobility.[18] Stealthy, long-range patrols can also be important in counterinsurgency operations and border security missions. Finally, noncombat operations, such as logistics, maintenance, and medical evacuation need to be part of the test situations.

Irregular warfare and counterinsurgency operations also affect the type of terrain and feature representation required for modeling the scenarios. Digital terrain elevation data level 2 terrain, produced by the Defense Mapping Agency with 30-meter elevation post-resolution,[19] was typically sufficient for mounted combat in open areas, but the more recent emphasis on dismounted warfare in rough terrain requires much higher resolution. There is substantial LIDAR (light detection and ranging)–based 1-meter resolution terrain available for Afghanistan.

---

[18] For descriptions of the battles of Wanat and Combat Outpost Keating, see Thomas E. Ricks, "Inside an Afghan Battle Gone Wrong: What Happened at Wanat?" *Foreign Policy*, January 28, 2009, and Michelle Tan, "Action Taken in COP Keating Attack," *Army Times*, February 11, 2010.

[19] Thirty-meter resolution means that an elevation measurement is provided at each 30-meter horizontal increment. The altitude between these posts is simply interpolated.

This level of detail is needed due to the typical target types that must be represented: people, animals, personal weapons, IEDs, and so on. In addition, a full description for the scenario (seldom included in AoAs or other study summaries) should include trafficability, cloud ceiling, line of sight, cultural alliances by region, and many other factors.

Stryker is a good example of the problems that can result from insufficient testing against a wide range of scenarios. Increasing use of IEDs, ambushes with rocket-propelled grenades, and operations in urban areas all highlighted the shortcomings of protection, vehicle weight, and off-road capability. Numerous rounds of testing and recapitalization were required throughout the program, and even when the Stryker was tested against the venerable M-113 under these conditions, the advantages were not overwhelming.[20]

Do We Have the M&S Tools to Answer the Questions?

In this section, we try to answer two questions: Do we have the models, simulations, procedures, and representations to adequately assess future ground vehicle designs? and What improvements to the set of tools are needed?

Currently, there is a wide range of analysis tools specifically for exploring and assessing military systems. They include engineering models showing, for example, how munitions penetrate armor and how suspensions react to obstacles at speed. They also include constructive force-on-force simulations, such as Combat XXI, in which abstractions of vehicles, soldiers, and weapons move, see, and shoot at each other on a digital map. Virtual environments take these force-on-force tools a step further, adding a dose of immersive, 3D virtual reality to the engagements. They are, by nature, more human-interactive and time-intensive to run than constructive models. The final category of tools contains everything else—those specialty models that calculate such aspects as fuel usage, maintenance events, logistics throughput,

---

[20] Thomas K. Adams, *The Army After Next: The First Postindustrial Army*, Stanford, Calif.: Stanford University Press, 2008.

and strategic deployment time. Also in the specialty model set is portfolio analysis, in which mixes of systems are optimized with respect to some cost criterion, such as manpower, dollars, or bandwidth.

It is even possible to link specific gaps and requirements for ground vehicles to specific tools. An example of this mapping of M&S tools to issues is shown in Table 5.2. The listing is by no means comprehensive; it provides an idea of the large number of models and procedures available to the analyst.[21]

Of course, no single tool or simulation can answer all of the analyst's questions and, in fact, each type of model or simulation has serious shortcomings in its application. Detailed engineering models, for example, tend to look at a very specific function (e.g., penetration of armor, overpressure effects, suspension travel, electronic countermeasure interference) and rarely take into account the other effects of a given change, such as increased weight (associated with the APS, for example) that impairs the vehicle's handling. Constructive simulations used for capability analysis typically take into account more interactions than engineering models but have limitations in their realism and responsiveness to an adaptive opponent. This is due to their largely scripted and rule-based nature. Virtual environments are much more responsive—they are driven in real time by a thinking human operator—but they are very burdensome in terms of time and manpower. Specialty models, such as logistics and maintenance analysis tools, deployment models, sensitivity analysis programs, and optimization tools, all tend to be used separately from the other models and may have inconsistent assumptions or use completely different scenarios.

There is much controversy over the use of Combat XXI, a constructive simulation, as the primary tool for AoAs and other key evalu-

---

[21] For a description of many of these tools, see Linda Kimball, "Modeling and Simulation Tools to Support the Challenges of T&E in an Urban Environment," briefing, U.S. Army Materiel Systems Analysis Activity, Huntsville, Ala., June 15, 2010; John Matsumura, Randall Steeb, John Gordon IV, Thomas J. Herbert, Russell W. Glenn, and Paul Steinberg, *Lightning Over Water: Sharpening America's Light Forces for Rapid Reaction Missions*, Santa Monica, Calif.: RAND Corporation, MR-1196-A/OSD, 2000; and Russell W. Glenn, Jody Jacobs, Brian Nichiporuk, Christopher Paul, Barbara Raymond, Randall Steeb, and Harry J. Thie, *Preparing for the Proven Inevitable: An Urban Operations Training Strategy for America's Joint Force*, Santa Monica, Calif.: RAND Corporation, MG-439-OSD/JFCOM, 2006.

Table 5.2
A Sampling of Specific Issues and Available Models

Focus of Analysis	Available Models
Digital terrain, features	ARC-GIS, Falcon View, constructive and virtual simulations
Tactical engagement modeling	Janus, JCATS (Joint Conflict and Tactical Simulation), CASTFOREM (Combined Arms and Support Task Force Evaluation Model), Combat XXI, OneSAF
Network modeling	Commercial models: QUALNet, OPNET
Jamming and interoperability	Skolnick radar equations, RJARS (RAND Jamming Aircraft and Radar Simulation), radio-frequency field tests
Human workload	Delta-3D, OLIVE (On-Line Interactive Virtual Environment), America's Army, OneSAF, field tests
Optimization	Portfolio Analysis Tool, PortMan
Logistics	Army log models: FMECA (Failure Modes Effects and Criticality Analysis), LORA (Level of Repair Analysis)
Maintenance	Army model for repair: COMPASS (Computerized Optimization Model for Predicting and Analyzing Support Structures)
Threat	STAR (System Threat Reduction Report)
New technologies	Specialty routines for Army prepositioned stock, acoustic sensors, stealth, precision munitions

ations of ground vehicle options. This model is an entity-based, force-on-force simulation with behaviors (moving, detecting, shooting, communicating) based on preset rules of action. These scripted actions play out in a vignette, and a single set of conditions (e.g., threat, terrain, weather) can result in a given distribution of outcomes, depending on the probabilistic (Monte Carlo) draws in the model. The advantage of this model is that a set of conditions can be quickly examined, and a wide range of runs completed quickly. The rules and behaviors have been validated over time, and the system has been accredited through observation by many SMEs.

The disadvantages of this model are those shared by many constructive models, such as Janus, JCATS, and CASTFOREM. All of these models were designed for major combat operations. As a result, most of the algorithms are concerned with movement, detection, and engagement of vehicles in open terrain. As operations have shifted to irregular warfare in complex terrain, the types of scenarios have become smaller in scale, with shorter lines of sight, increased presence of noncombatants and infrastructure, and greater use of cover and concealment. At the same time, the dynamics of the combat situations have changed, from long time lines during movement to contact and good situational awareness to ambushes with little or no warning, as well as IED attacks and sniper fires. This has shifted the modeling requirements to simulations and tools that can more accurately capture the speed, uncertainty, and adaptation of irregular warfare and urban operations. Largely scripted constructive simulations are also not well suited to measuring human workload or exploring the robustness to enemy countermeasures. Figure 5.5 shows a screenshot of a constructive simulation, Janus, with its map-type display of digital terrain and entities.

Virtual environments, such as those embodied in the newer gaming simulations (OLIVE, Delta-3D, Unreal Tournament, and many others), may be more appropriate for these new modeling challenges. Figure 5.6 shows some screen images from one of these environments, the open-source Delta-3D system. Virtual environments have the advantage of being able to immerse the user in the situation and seeing what behaviors emerge. This provides a much more realistic and flexible backdrop for evaluating systems than constructive simulations but tends to be time- and manpower-intensive. It is also difficult to compare options because the human users learn from iteration to iteration.

A compromise may be the OneSAF system, which has characteristics of both constructive simulations and virtual environments. The semiautomated forces component can model behaviors autonomously (without operator input during the run), using rule sets, while the virtual component allows users to interact with elements in the battle, often with semirealistic controls. Both the autonomous and interactive components can act simultaneously, allowing large battles to be fought with limited numbers of SMEs.

Figure 5.5
Screenshot of Force-on-Force Constructive
Simulation (Janus) Representing a Convoy
Operation in an Urban Area

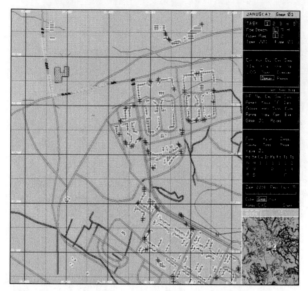

SOURCE: Randall Steeb, John Matsumura, Paul Steinberg,
Thomas J. Herbert, Phyllis Kantar, and Patrick Bogue,
Examining the Army's Future Warrior: Force-on-Force
Simulation of Candidate Technologies, Santa Monica, Calif.:
RAND Corporation, MG-140-A, 2004, p. 46, Figure 5.7.
RAND *MG1093-5.5*

As noted earlier, virtual environments, such as OneSAF with
human-in-the-loop modeling, also have many disadvantages. They are
often expensive, time-consuming, and difficult to control. Because of
their complexity, they can be vexing to verify, validate, and accredit.
These problems are becoming less burdensome with time, primarily
due to improvements from the gaming industry (with the development
of standardized libraries of models, postprocessing algorithms, and so
on). Nevertheless, virtual environments are unlikely to take over the
bulk of the analysis process without substantial investment.

Over the years, many attempts have been made to develop a
linked federation of models that can represent all echelons and functions, from individual soldiers and subsystems up to brigade or division.

Figure 5.6
Realistic Images Generated from Delta-3D

SOURCE: Moves Institute, Naval Postgraduate School.
RAND *MG1093-5.6*

Some of these, such as WARSIM (Warfighters' Simulation), JWARS (Joint Warfare System), JSIMS (Joint Simulation System), and the FCS System of Systems Simulation Environment, combine constructive and virtual simulations. Some large experiments, such as Millenium Challenge, Urban Resolve, and Army After Next, also add live actors to the mix.[22] Workshop participants noted that all these efforts have suffered from problems with synchronization of models, incompatible interface protocols, speed difficulties, and other integration issues. It would appear that a few well-chosen tools, loosely connected, may be more tractable at this juncture.

---

[22] For descriptions of these systems, see Glenn F. Gutting, "Inside Urban Resolve 2015 and Omni Fusion 2006," TRADOC News Service, October 25, 2006, and Walter L. Perry, Bruce R. Pirnie, and John Gordon IV, *Issues Raised During the 1998 Army After Next Spring Wargame*, Santa Monica, Calif.: RAND Corporation, MR-1023-A, 1999.

Exploring the Trade Space

Probably the most important role of M&S is in performing trade-offs among key factors, such as protection, situational awareness, mobility, and firepower. In this section, we discuss some of the tools for performing such trade-off analyses and illustrate these actions with some examples from recent studies. Some of the more important trades for ground combat and tactical wheeled vehicles are as follows:

- protection versus mobility
- passive versus active protection
- manned versus unmanned systems
- armor versus situational awareness
- payload versus speed versus range
- range versus off-road capability
- size versus stealth
- modularity versus specialization
- organic ISR and fire support versus remote.

These trades are especially important for ground vehicles because the platforms are pushing against the limits of weight, power, size, maneuverability, payload, protection, and cost. At the same time, there are also issues of unintended consequences of multiple systems working together. This was seen, for example, when the introduction of CREW (counter–radio-controlled IED electronic warfare) produced interference with network communications.

The process of establishing trade-offs with future ground vehicles, as with any new technology, is not a simple matter. First, the requirements or needs must be translated into measures of effectiveness and measures of performance. These should encompass both individual system measures and those for the force as a whole. Evaluation of options should then be made with a variety of models, with an emphasis on the use of multiple options being evaluated both in isolation and in concert, to see whether unique contributions, synergies, or conflicts arise. The performance of each option then needs to be examined to see whether there are thresholds of minimum performance, flat maxima, inflection points, or other instances in which trades may be called for. Finally, sensitivity analyses need to be conducted with options in isolation and with options in combination. This type of analysis is extremely difficult to accomplish without M&S. Simple computations will seldom show the complex interactions between options; field tests may show these interactions, but the number and cost of the excursions would likely be prohibitive.

One problem is that constructive simulation, with its emphasis on repeatable scripted actions across multiple conditions, will not be adequate for modeling complex trade-offs. In fact, some shifts, such as moving from a 30-ton platform to a 16-ton one, may facilitate completely different tactics (e.g., air-mechanized operations). In a similar manner, the use of unmanned systems may result in the development of high-risk, revolutionary tactics and a reduced emphasis on protection with all that implies, rather than simply recreating the actions of manned systems.

Trade-off analysis was critical in the decisions leading to the refinement and even the demise of the FCS program. Use of M&S showed shortcomings in vehicle vulnerability, strategic and operational mobility, airlift survivability, weapon redundancy, and level of situational awareness in complex terrain. Many of the following issues were highlighted or quantified by M&S:[23]

- Vehicle weight was critical. Light vehicles were vulnerable to fires but enabled air-mechanized capability.
- The APS had limitations. It protected against low-speed ATGMs but not against multiple simultaneous fires and kinetic rounds.

---

[23] See Paul L. Francis, Director of Acquisition and Sourcing Management, U.S. General Accounting Office, *Defense Acquisitions: The Army's Future Combat Systems Features, Risks, and Alternatives*, testimony before the House Committee on Armed Services, Subcommittee on Tactical Air and Land Forces, Washington, D.C.: U.S. General Accounting Office, GAO-04-635T, April 1, 2004, and John Matsumura, Randall Steeb, Thomas J. Herbert, John Gordon IV, Carl Rhodes, Russell W. Glenn, Michael Barbero, Frederick J. Gellert, Phyllis Kantar, Gail Halverson, Robert Cochran, and Paul Steinberg, *Exploring Advanced Technologies for the Future Combat Systems Program*, Santa Monica, Calif.: RAND Corporation, MR-1332-A, 2002.

- Precision weapons were a force multiplier, but several of the systems were found to be redundant and costly.
- Comprehensive situational awareness was seen to be important in open areas but could not be achieved in urban operations and irregular warfare.

On the positive side, M&S did show the significant value of FCS air-mechanized operations against an entrenched opponent, provided that the overflight was not in areas with dense air defense systems. Some of these findings were determined from computational analysis, while others were found from constructive and virtual simulation.

Several recent RAND studies also showed the importance of M&S for evaluation of new concepts and how the assessments could lead to refinement of tactics and the selection of technologies. These studies showed the interdependency of ground and air vehicles, the importance of sensor height for situational awareness, the limitations of indirect fires against an enemy in complex terrain, and the problems with the APS against simultaneous fires.[24] These many studies illustrate several points that were alluded to earlier: (1) the choice of scenarios and vignettes is very important (nonstressing conditions rarely discriminate between alternatives), (2) richness of conditions is critical to testing system robustness and flexibility, and (3) more often than not, the preparation of the analysis (e.g., terrain formatting, laydown of forces, timing of actions, establishing a baseline) is more demanding than the actual analysis.

Once the attributes of various systems are defined through force-on-force analysis and other modeling, tools are needed to help the decisionmaker determine the best mix and organization of platforms and systems. This can be done using sensitivity analysis, dynamic constraint satisfaction, portfolio analysis, and other satisficing or optimiza-

---

[24] See the following RAND publications: Matsumura et al., 2000; Matsumura et al., *Exploring Advanced Technologies for the Future Combat Systems Program*, Santa Monica, Calif.: RAND Corporation, MR-1332-A, 2002; Steeb et al., 2004; and Randall Steeb et al., *A Simulation-Based Exploration of Options to Protect Small Units, Based on a Case Study of the Battle of Wanat, Afghanistan*, Santa Monica, Calif.: RAND Corporation, 2010, not available to the general public.

tion tools. Some notable examples are the PortMan portfolio analysis decision framework and the Portfolio Analysis Tool.[25] These tools are intended to assist decisionmakers by providing a transparent and auditable process for balancing programs across a series of defined metrics to achieve a specific capability or set of capabilities.

Providing the Decisionmaker with the Background Needed to Properly Select M&S Tools and Supervise the Process

A key outcome of the project workshop was the notion that the decisionmaker is often not fully cognizant of the best use of analysis tools and does not understand all the insights and limitations of M&S. Some of this is due to a need provide the decisionmaker with the needed background and information to properly select M&S approaches, and some is a result of models that are not sufficiently transparent in their assumptions and behavior.

Discussions with M&S professionals and program decisionmakers have led us to the conclusion that many misunderstandings concern the appropriateness, performance, and costs and burdens of the different types of analysis tools. Decisionmakers often request to use very sophisticated tools during the early conceptual phases of development and during the later production phases, both of which are periods when these tools can provide only limited information and direction. There are also some misconceptions about which tools should be used in the training community and which should form the basis for analysis.

If we tried to plot the various models in terms of their cost (e.g., in time and manpower, as discussed earlier) and their level of detail and discrimination, we might arrive at something like Figure 5.7. The easiest but least definitive tools are SME war games, which are extremely useful early in the development cycle, while the most difficult and

---

[25] For an example of PortMan in naval applications, see Richard Silberglitt, Lance Sherry, Carolyn Wong, Michael S. Tseng, Emile Ettedgui, Aaron Watts, and Geoffrey Strothard, *Portfolio Analysis and Management for Naval Research and Development*, Santa Monica, Calif.: RAND Corporation, MG-271-NAVY, 2004.

Figure 5.7
A Rough Plot of Tool Cost and Level of Detail for Capability
Development

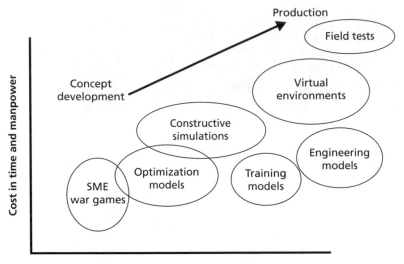

RAND *MG1093-5.7*

discerning tool is the physical field test, best employed near the end of development, when prototypes are available and operational TTPs have been at least initially defined. In between, there are many overlapping areas where multiple models are appropriate for a given phase of development or acquisition.

At the same time, decisionmakers would likely have a better appreciation of what these models can and cannot do if the models were more transparent in their assumptions and behavior. Most studies are not sufficiently characterized (or reported) in terms of data assumptions, conditions, entity behaviors, communications, and actions that led to specific outcomes. A shift from the more opaque constructive simulations to more transparent and immersive virtual environments should help, both during the actual runs and in the after-action analysis. The visual replay nature of virtual environments should also allow greater understanding and questioning of assumptions by the decisionmaker. This should lead to more interchange among decisionmakers, users, developers, and analysts and an increased confidence in the product.

Of course, the services and institutions often have their own issues with insularity and lack of commonality and interoperability. Development, analysis, and testing commands (e.g., TRADOC, TRAC, Army Test and Evaluation Command, RDECOM) typically have their own favorite models, scenarios, and post-processing tools. This makes it difficult to compare results across institutions and introduces redundancies and friction points. Of course, some specialized simulations, such as spall, fragmentation, and blast models, will be specific to certain commands, but the more general force-on-force models (and their associated VVA) should be common to all developers and users.

The Army After Next war game in 1998 was an instructive example of misunderstandings among decisionmakers, users, analysts, and developers. The intent of this major exercise was to explore lightweight, highly deployable options for ground forces. Many of the participants assumed that network-based situational awareness could overcome protection issues and that rapid deployment could provide major tactical advantages. The resulting effects-based operations fared well against a traditional opponent, but the array of high-tech systems were nowhere near as effective against an insurgency.[26] However, this realization was slow to emerge from the proceedings.

With improved education of decisionmakers, greater transparency of the modeling process, and increased use of common models and procedures, there should be more efficient use of resources and greater confidence in and agreement with the findings.

Summary and Investment Recommendations

The primary challenge to M&S for capability development and evaluation in DoD appears to be in maintaining the agility to keep up with changing battlefield requirements and ensuring development of new

---

[26] Adams, 2008.

tactics and technologies. Among the trends we have seen that require a more streamlined and efficient analysis process are the following:

- Nontraditional acquisition processes (e.g., rapid-equipping force, urgent needs requests) often require accelerated scenario development, modeling, and experimentation cycles.
- Modularity of electronic architectures means that subcomponents could be added to any vehicle type; M&S will need to be able to quickly add subroutines for each added capability.
- Trade-offs of overlapping protection devices (e.g., APS, electric armor, appliqué armor) mean that M&S must be able to quickly represent both the new capability and the impact on space claims, mobility, and power.
- Sophisticated network development implies improvements in situational awareness at many levels; models will need to better represent special functions for electronic warfare, communication jamming, and interoperability.

All these challenges color the ways in which M&S can respond to the congressional concerns regarding ground combat and tactical wheeled vehicles, as enumerated in Chapter One. These concerns were, briefly, discussion of requirements and capability needs, identification of capability gaps, determination of critical technology elements or integration risks, and recommendations of actions to develop and deploy critical technology capabilities. While we cannot, in the short time allotted to the study, make definite, analytically supported recommendations to confirm or deny currently identified gaps, requirements, capability needs, and critical technology elements, we can make a set of recommendations for investment in and redirection of the current M&S process.

Most of our recommendations deal with changes that the commands must make to streamline the M&S processes dealing with scenarios, tools, and institutional exchange:

- Scenario development is too slow and diffuse, resulting in vignettes that are often insufficiently stressing and quickly obsolete.

- The level of detail of the supporting digital terrain is often too coarse for irregular warfare.
- Models need to be updated, moving from a reliance on opaque constructive simulations toward more transparent and agile virtual environments. At least, there should be greater investment in systems, such as OneSAF, that have both constructive and virtual components. If OneSAF and other virtual environments are to be used more widely, greater investments need to be made in VVA of these tools.
- Once the force-on-force analysis process is running smoothly, specialty models need to be more closely integrated. In particular, new optimization tools need to be incorporated in the final stages to help develop operational architectures.
- Common tools and assumptions need to be shared among development, analysis, and testing commands.

Finally, the M&S process itself can be improved by setting up standard procedures for integrating all the various development and acquisition tracks: deliberate, nontraditional, recapitalization, and even common-module insertion. This consolidation will require the participation of everyone involved—users, developers, analysts, testers, contractors, and decisionmakers—and will necessitate a common language for establishing measures of effectiveness and measures of performance, scenarios, models, and procedures.

CHAPTER SIX

Observations and Conclusions

The observations in this chapter follow the basic structure of the analysis contained in this monograph, focusing on technology-based issues, policy- and business process–related insights, and M&S and concluding with general observations about what is working and not working. These observations attempt to weave together the individual insights in a way that answers the congressional questions that prompted this study: the specific issues addressed in the legislation and, in particular, whether the future combat and tactical wheeled vehicle fleet is likely to be suitable for the anticipated future operating circumstances.

Requirements-Related Issues

In our interactions with combat developers from the Army and Marine Corps, we found no evidence of fundamental flaws in their requirements development processes, though we did not independently develop our own requirements to validate DoD's, nor did we examine these processes in depth. We were able to observe that arriving at a satisfactory set of requirements for tactical wheeled and ground combat vehicles is complicated by the fact that the vehicles remain in the services' inventories for decades. It is difficult to anticipate the circumstances in which they will be employed over such an extended period (e.g., the enemies and weapons that may confront them, the weather in which they must operate, the terrain they must traverse).

Combat developers typically have a deep understanding of today's and near-term future operating requirements but cannot unfailingly predict the future. Intelligence Community threat documents, while certainly important to understanding the most likely and most stressing enemies and weapons that may confront U.S. military vehicles, cannot predict all threats associated with the future either.

The implications of these circumstances are that, all but inevitably, *DoD will have vehicles in its fleets that were designed and built for requirements that differ somewhat from those it finds itself facing in the future.* This fact is driven by the wide spectrum of potential threats and scenarios in the 21st century and the fundamentally different physics and engineering problems presented by conflict with these threats. It is not possible to design vehicles that are optimal for all possible scenarios. Choices will have to be made that consider several aspects of operational risk against a portfolio of scenarios, seeking to create vehicles that minimize risk across a number of different factors, rather than minimizing risk for a single, well-defined threat and scenario (as was possible during the Cold War). Moreover, weapons—be they IEDs or new rockets—evolve quickly and cheaply or are fielded before defenses against their effects can be developed. Furthermore, it is no longer reasonable to expect that on-vehicle protective capabilities will be sufficient to defend a vehicle and its occupants against all threats.

The full set of desired operational requirements is unlikely to be met in many cases. Because of the constraints on the trade space into which all vehicle requirements must fit, the resulting vehicles are unlikely to deliver 100-percent performance against all desired design criteria. In particular, program requirements will not meet all operational requirements. As noted several times in the preceding chapters, vehicle weight drives trade-offs with mobility, protection and survivability, strategic mobility, and logistical requirements (e.g., fuel consumption and maintenance), to name some of the most important.

The iron triangle of trade-offs is permanent. In particular, DoD will always want vehicles that provide better protection, have more power (electrical and mechanical), and perform better or are more capable (e.g., weight, mobility). No matter what technical advances are made, there will always be a drive to do better in these categories because advances will help protect soldiers and marines; make the U.S. military more mobile strategically, operationally, and tactically; and increase performance. In other words, the technological and engineering challenges presented by the iron triangle (performance, protection, and payload) are permanent. However, advances in networking (not explicitly considered in this study) and other technologies can, in part, move some of these challenges to the system of systems on the battlefield, rather than remaining challenges for vehicle design only. In particular, protection is a unit and joint force mission not limited solely to individual vehicle performance. Specifics related to these trade-offs are discussed later.

The vehicles resulting from this process may fail to meet all requirements but may nevertheless be satisfactory in the sense that they are sufficiently better than existing vehicles to make them worth the investment. Even if they fail to deliver the full suite of capabilities envisioned by operational requirements or even program requirements documents, they may be superior to the vehicles they replace (e.g., the JLTV versus the HMMWV). If they are affordable, perform reasonably well, and allow the U.S. military to prevail in its operations, they may be said to be satisfactory because vehicles were lost or underperformed at a rate that did not ultimately compromise U.S. military operations. Most importantly, there may be no better alternatives.

These observations with respect to requirements have implications for technology- and engineering-related issues, as well as for acquisition-related processes. The technology- and engineering-related issues most closely align with the questions asked by Congress in Section 222 of the National Defense Authorization Act for Fiscal Year 2010; however, they will be much more likely to come to fruition at a reasonable cost and within a reasonable time frame if the acquisition process–related issues are also addressed.

Technology-Related Issues

The study found four major technology-related issues associated with the vehicle fleets to be the most challenging with respect to meeting operational requirements. They were protection, power generation, fuels and fuel consumption, and sensors, networking, and complexity. We treat each issue in turn.

Protection

The critical observations with respect to protection are as follows:

- Protection requirements differ based on expected threats, and technical and engineering solutions will differ based on these requirements.
- Protection requirements should consider onboard and offboard technology, as well as vehicle design and integration improvements.
- Improving protection will be a permanent task to which technology and engineering will need to contribute (along with tactics, unit designs, and other factors); it will never be good enough.

The first two observations have some common technical implications. They include the need for continued research in armor, particularly materials that are stronger and lighter; configuration or design of vehicles that can mitigate competing requirements derived from different potential threats beyond just making stronger, lighter vehicles (e.g., the newly discovered "stovepipe" design that permits the blast effects of underbody IEDs to pass through a vehicle may permit future vehicles to maintain lower profiles—desired for direct-fire threats—while maintaining robust protection against IEDs);[1] and efforts to enable protection against most major categories of threats, across a broad spectrum of conflict scenarios, through enhanced situational awareness (a technical task) and improved unit designs (a task whose success would rely heavily on M&S).

We note as well that there are significant procedural and organizational innovations that would enable these technologies to enhance protection (as well as the other technical tasks). These innovations are discussed later in this chapter.

---

[1]   Kate Brannen, "'Chimney' Deflects IEDs: Humvees Survive 3 Aberdeen Tests," *DefenseNews,* November 1, 2010g.

Electrical Power Generation

The last generation of tactical wheeled vehicles did not differ significantly from its civilian counterparts as far as electrical systems were concerned. The majority had simple 24-volt electrical systems that were generally sufficient to start their engines; power their headlights, turn signals, and blackout drive lights; and support a single voice radio. Ground combat vehicles were somewhat more power-hungry, but only at the margins; their electrical systems typically had to be adequate to support thermal imaging sights, computer displays, intercoms, and multiple radios. In circumstances in which more electrical power was required, units deployed generators to provide it for command posts, radars, and similar applications.

The advent of tactical networks, computer-based battle command systems, and expectations of battle command on the move (that is, without stopping to set up antennas and generators) changed the demand for electrical power. The tactical network produces situational awareness, answering soldiers' and marines' fundamental battlefield questions: Where am I? Where are my buddies? Where is the enemy? The network not only shares a common picture of the battlespace but also identifies and keeps track of friendly units through Blue Force Tracker, a system mounted on both tactical wheeled and ground combat vehicles that reports the vehicle's location periodically. Sustainment units also deploy inventory visibility systems. Combat units operate battle command systems. These systems are radio- and computer-based and drive demand for electrical power upward. In addition, counter-IED equipment is mounted on vehicles of all types and requires significant amounts of power.

In some instances, fitting larger alternators onto the vehicles to supply the necessary power can satisfy the additional demand for electrical power. In other circumstances (e.g., reconnaissance, surveillance, and target acquisition squadrons and reconnaissance vehicles), the combination of long duty cycles, which require the vehicles to operate nearly continuously, and the need for "silent watch," in which a reconnaissance vehicle must operate its sights and sensors but the engine must be shut off to avoid presenting the enemy with noise and thermal signatures, outstrip the ability of current systems to provide the requisite power. Under these circumstances, the vehicles require large battery storage, fuel cells, or auxiliary power units to provide the necessary electricity.

In addition to these on-vehicle considerations, reducing the need for external generators and associated equipment and support enhances strategic and operational mobility and potentially reduces logistical requirements. These are important considerations for the Marine Corps, in particular, given its expeditionary charter.

The network, battle command systems, inventory visibility systems, counter-IED jammers, and reconnaissance, surveillance, and target acquisition sensors are all evolving; the ultimate need for electrical power for these and other future electronic pieces of equipment cannot be accurately anticipated. Every new piece of equipment related to these systems comes with a need for electricity, a need for space on the vehicle where it can be installed and operate without interfering with other vehicle functions, and heat-management requirements (that is, a way to manage heat created by the equipment so that the vehicle's interior does not become hot enough to interfere with the functioning of onboard electronic systems or degrade crew performance). Thus, the demand for additional electrical power means that not only must the vehicles be able to provide the electricity but vehicle designs must be such that they can accommodate the space, weight, and cooling needs associated with the additional equipment. It is also important that the vehicle's space, weight, power, and cooling capabilities are flexible enough to accommodate new equipment that evolves later.

The demand for additional electricity affects the designs of both tactical and combat vehicles. Tactical wheeled vehicles must increasingly be able to support network systems because sustainment units and other noncombat formations operate in the same noncontiguous areas of operation as their combat counterparts and therefore need the same quality of situational awareness in order to survive and accomplish their missions. The designs of tactical wheeled vehicles have therefore become more complicated and the unit costs have grown relative to the vehicles they replaced. The up-armored HMMWV, for example, is a basic utility vehicle with a per-unit cost of approximately $186,000.

The JLTV, its network-ready replacement, is estimated to have a per-unit cost of $306,000–$332,000 (depending on the version).

Growing demands for additional electricity affect future tactical and combat vehicles in important ways (see Figure 6.1). Currently, future capabilities have not been fielded or field-tested on vehicles but should come to fruition in the time frames that Congress is considering. Making this possible will be a critical engineering effort.

Future vehicles will almost certainly be more expensive than their predecessors, because they will need advanced power generating and other capabilities. Moreover, their designs must have "open architecture" to accommodate future network-related equipment, along with the additional weight and space that this equipment will claim on the vehicle and the heat the new components will generate. Vehicle designs will therefore become more complex, and complexity will drive unit cost upward. As tactical wheeled vehicles come to have progressively

Figure 6.1
Power Demands and Developments

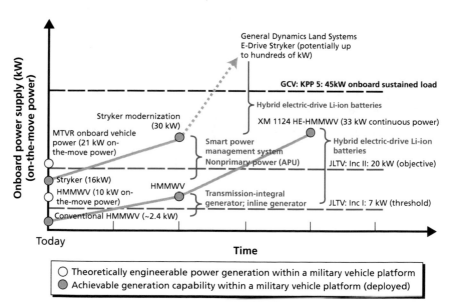

NOTE: APU = auxiliary power unit.
RAND *MG1093-6.1*

less in common with their civilian counterparts, they will become more expensive to produce because building them will require more effort than simply modifying civilian vehicles for military applications.

Figure 6.1 illustrates the proposed requirements, along with some of the potential technology solutions for meeting them. The dashed lines identify the threshold (increment 1) and the objective (increment 2) requirements for the JLTV and the GCV KPPs for onboard vehicle power mentioned in Chapter Four. There have already been developments in transmission-integral generators that have enabled HMMWVs to generate up to 30 kW of power while stationary and 10 kW of power while on the move.[2] According to DRS Technologies, this same technology is capable of producing up to 125 kW of power (stationary) in larger combat vehicles, such as the MRAP and the Stryker.[3] These types of technologies may enable the JLTV to meet some of its threshold (and possibly objective) requirements for onboard vehicle power.

Similarly, as part of the Stryker modernization program, plans are to improve the onboard vehicle power to 30 kW, from its current limit of 16 kW. Technologies being considered to help the Stryker activate the desired onboard vehicle power include smart power management systems and additional (nonprimary) auxiliary power units. However, looking even further into the future, hybrid-electric systems may be necessary to achieve the desired vehicle power requirements. However, many important research challenges persist, including developing and integrating silicon-carbide electronics to increase efficiency and reduce power loss, improving lithium-ion batteries to enable increased capacity and higher thermal operating temperatures, and improvements in ultracapacitors, which will enable improved regenerative braking capabilities.

The flip side of power generation is the efficiency of those items that place additional power demands on vehicles. The need to develop

---

[2]   DRS Technologies, 2010b.

[3]   DRS Technologies, "DRS Technologies Provides Critical Power Technology to United States Marine Corps for On-the-Move Power Needs of Tactical-Wheeled Vehicles," news release, June 28, 2010a.

new ways to generate power is a function of demand. To the extent
that innovations can reduce power demands, the generation requirement becomes less challenging. Furthermore, other innovations, such
as improved batteries, can also help ease the power-generation requirements by cutting down on duty cycles and permitting a greater per-
centage of the demanded power to be produced when vehicles would
otherwise be operating.[4]

Fuels and Fuel Consumption

Fuel costs and availability are major factors in ongoing and possible
future Marine Corps and Army operations. Some of the officials with
whom we spoke estimated that the fully burdened cost of a gallon of
fuel (that is, its cost including transportation charges) in remote areas
of Afghanistan exceeds $900 per gallon. The same officials indicated
that U.S. forces in Afghanistan consume more than 1 million gallons
per day.[5] Many of the convoys that must face the dangers involved in
supplying remote combat outposts are necessary because they deliver
fuel. Therefore, the ability to burn indigenous fuels (including high-
sulfur, "dirty" fuel) and fuel-efficient engines that deliver adequate
power are critical to reducing operating costs and human losses associated with hauling fuel during combat operations.

Future conflicts could pose even more challenges with respect to
fuel. It is possible that in some future scenario, U.S. forces would be
unable to secure enough fuel from international supply routes, forcing
them to depend on local fuels (which at the moment they cannot use
without damaging some equipment).

DoD is conducting efforts to improve the fuel efficiency of its
vehicles. This is a major focus for the Marine Corps, and the Army's

---

[4] See U.S. Army Research, Development, and Engineering Command and U.S. Army Tank
Automotive Research, Development, and Engineering Center, "JLTV Briefing to Industry,"
May 2009, for a brief overview of battery developments and other power-related issues.

[5] According to one report, the Marine Corps alone consumes as much as 800,000 gallons
of fuel per day (see Tiron, 2009). This is not meant to imply that fuel costs in Afghanistan
are $900 million per day; they are not.

RDECOM-TARDEC is also considering fuel efficiency as part of its Ground Vehicle Power and Mobility Program.[6]

Sensors, Networking, and Complexity

Sensors and networks are outside the formal purview of this study, but due to their significant effects on many aspects of vehicle design and modification, ranging from protection to power to space and cooling considerations, how these functions develop and are implemented must be briefly considered. They are critical technologies that will be important for Congress to consider.

The earlier discussion of requirements for electrical power generation touched on most of the salient dimensions of this issue. One aspect of sensors and networking that has not been dealt with in these observations is complexity. Complexity grows as vehicles or systems gain operating attributes—in this case, more sensors, connectivity, throughput, nodes, and so on. Complexity can create the conditions in which unidentified dependencies and incompatibilities among components and subsystems can result in performance problems, sometimes posing significant slowdowns in the systems integration effort. A potential outcome is schedule slippage and cost growth. Hedging against the effects of complexity requires additional efforts in systems engineering and systems integration. In addition, a nontrivial potential exists for complexity to mask deficiencies that can affect vehicle performance.

Because sensors and networking add complexity to vehicle designs, there is a greater chance for schedule slippage and cost growth with the vehicles currently under development than there was with their simpler predecessors. Moreover, if our hypothesis is correct that tactical wheeled vehicles will increasingly need network and sensor functionality similar to that of ground combat vehicles to support their situational awareness, counter-IED, and communication capabilities—and, therefore, their survivability and mission accomplishment—complexity is likely

---

[6]   See U.S. Army Research, Development, and Engineering Command and U.S. Army Tank Automotive Research, Development, and Engineering Center, 2009.

to emerge in the development of tactical wheeled vehicles as well as ground combat vehicles.

Complexity is a challenge to DoD's ability to field an adequate vehicle fleet only insofar as it affects operational or program success. Indeed, increased complexity is often the result of efforts to develop enhanced capabilities—that is, greater operational capabilities—and so better meet operational requirements. To deal with the challenges posed by complexity and the ability to understand it, the services may require both more systems engineering and systems integration capability to manage complexity, as well as more research and development to understand emerging technical challenges.

Acquisition Policy– and Business Process–Related Issues

At least seven key observations based on prevailing DoD policies and business processes bear on the services' ability to field vehicles appropriate for the anticipated operating circumstances. These include (1) the funding implications of gradual merging of some key performance criteria of tactical wheeled and ground combat vehicles, (2) the lack of stable funding and requirements for programs developing and buying new systems, (3) modified cost-estimating procedures, (4) the lack of alignment of M&S with decisionmaking and the needs of decisionmakers, (5) program categorization into ACATs based on risk, (6) ensuring that programs are "born healthy" (i.e., have appropriate fiscal and human resources from the beginning), and (7) the need for a different approach to integrating test and evaluation into individual vehicle programs.

Funding Implications of the Merging of Some Key Performance Criteria for Tactical Wheeled and Ground Combat Vehicles

As pointed out earlier in this chapter, in the section "Protection," many of the tactical wheeled vehicles currently in the force were procured under assumptions that did not lead to substantial requirements for situational awareness and communication capabilities in those fleets. Those assumptions have been largely overturned by today's operations and replaced with a growing appreciation of how these capabilities can facilitate unit performance and mission accomplishment. As a result, tactical wheeled vehicles are acquiring more situational awareness and protection capabilities, thus growing closer to their ground combat vehicle cousins and more distant from their commercial counterparts. These trends mean more expensive vehicles in most fleets and, due to the very large number of tactical wheeled vehicles, much more expensive fleets. The trade-off between survivability and affordability presents a major policy decision for DoD and Congress.

Stable Funding and Vehicle Requirements

Many of the acquisition officials with whom we spoke believe that funding instability and creeping vehicle requirements are among the biggest threats to their programs.[7] There is a tendency, according to this view, to spread available funding across too many programs in order to sustain them all. For example, in 2008, the House Armed Services Committee expressed its concern that Army leaders had "dramatically reduced funding for [Abrams, Bradley, and Stryker] vehicles" and that this could lead to instability in, or termination of, vehicle programs.[8] Although this practice has the intended effect of providing life support to a larger number of programs than would otherwise be sustainable, it introduces uncertainty into program managers' planning and project management, with negative effects on the program schedule and budget. Creeping program requirements have similarly pernicious effects. The change order process is cumbersome and can delay program progress, add cost, and add complexity, with all that it implies.

---

[7] Funding can be unstable for a variety of reasons, including service or DoD funding decisions (as noted in the text), changes in program costs that have the same effect as changes in funding levels, and changes in congressional priorities. This monograph does not examine these causes in detail, but we do note that this was the most frequently stated concern of program managers with whom we spoke.

[8] "HASC Leaders Write Gates About Army Abrams, Bradley and Stryker Funding," *Defense News Daily*, October 1, 2008.

Cost-Estimating Procedures

Differing cost-estimating procedures yield more or less useful insights into the total costs associated with adding capabilities to vehicles, providing flexibility for future modifications, and similar decisions. Among the officials who commented on cost estimating, most believed that life-cycle estimates were superior to unit cost estimates to support such decisions. GAO reports verify this, though in a more general context.[9] In particular, there was general consensus among the acquisition personnel with whom we spoke that different acquisition decisions are made and net life-cycle costs reduced when cost estimates included these considerations.[10] Earlier research indicates that the services typi-

---

[9]   U.S. Government Accountability Office, *Defense Management: DoD Needs Better Information and Guidance to More Effectively Manage and Reduce Operating and Support Costs of Major Weapon Systems*, Washington, D.C., GAO-10-717, July 2010c, notes the following general problem with DoD cost estimates:

> DoD lacks key information needed to effectively manage and reduce O&S [operating and support] costs for most of the weapon systems GAO reviewed—including life-cycle O&S cost estimates and complete historical data on actual O&S costs. The services did not have life-cycle O&S cost estimates developed at the production milestone for five of the seven aviation systems GAO reviewed, and current DoD acquisition and cost-estimating guidance does not specifically address retaining these estimates. Also, the services' information systems designated for collecting data on actual O&S costs were incomplete, with the Army's system having the greatest limitations on available cost data. Without historic cost estimates and complete data on actual O&S costs, DoD officials do not have important information necessary for analyzing the rate of O&S cost growth for major systems, identifying cost drivers, and developing plans for managing and controlling these costs.

[10] The Office of Naval Research provided two examples of significant cost savings that could be achieved with technology insertions into existing platforms. MTVR fuel-efficiency upgrades look likely to be adopted by the Marine Corps and promise significant savings, with the savings dependent on the FBCF. However, a solicitation for advanced suspension systems on HMMWVs included target costs that were much lower than those required to field more advanced suspension systems, such as an electronically controlled suspension that demonstrated significant performance gains and life-cycle cost reductions over the current suspension. As a result, the contractor that built this technology demonstrator did not propose this system in response to the RFP, and so the system that the Marine Corps buys may not provide the same advantages (Office of Naval Research, "O&S/LCC Cost Reductions Through Technology Insertions," briefing, January 4, 2011, and email exchanges with Office of Naval Research personnel, January 2011).

cally underestimate life-cycle costs significantly. Better life-cycle cost estimates are needed, too.

Estimates often in use today seem to be driven by two factors, according to experts consulted by the RAND team for this work. First, the acquisition community (as with every other DoD community) has to operate within its own budget category, so it focuses primarily on the factors that affect program success. These factors may not adequately include such considerations as realistic operation and maintenance costs, for example. Second, because funding is allocated annually to DoD for all programs, decisions are made to maximize program success over short time horizons and are not based on what would provide the best value for the dollar over the lifetime of a system.[11]

There are indications that some of these concerns may be addressed through pending changes to acquisition practices.[12] These changes could include more comprehensive cost models. Alternately, some concerns could be captured in requirements documents that more carefully stipulate which operation and maintenance considerations are important to the services.

M&S Alignment with Decisions and Decisionmakers

One insight of particular importance with respect to M&S has to do with policy and business processes: M&S efforts do not appear to be fully aligned with the decisions they are meant to support (e.g., whether a materiel solution is warranted, technology development, AoA, milestone decisions) and the information needs of the officials who will make them. If the services are to enjoy the full benefits of the M&S conducted to assist the research, development, and support of vehicle programs, they must make a greater effort to perfect this alignment. Such an effort would involve canvassing the research and development community, milestone decision authorities, program executive officers, and program managers to identify the decisions they make that would benefit from M&S insights and what, specifically, those insights

---

[11] Comments made by senior DDR&E officials during a meeting with RAND researchers, September 2010.

[12] See the directions on how to consider cost estimates in Carter, 2010c.

would look like. Furthermore, better education for these officials on the costs and benefits of certain M&S approaches at different phases of the acquisition cycle could help DoD better use its M&S capabilities and save money.

ACAT Decisions Should Emphasize Risk Rather Than Cost

Currently, ACAT decisions are based on cost alone. Risk, the minimizing of which is the driving concept behind ACAT decisions, is not considered, except to the extent that cost is used as a proxy for risk. However, risk in this context is a function of the likelihood of failure and its consequences. Cost addresses only the latter argument of this function, and only in part. As a result, mature, well-understood, but expensive programs contemplating changes and modifications that pose little risk are nevertheless subjected to stringent requirements meant to manage risk. At the same time, other, relatively low-cost programs that embody significant risk may not receive the attention they would otherwise deserve because the overall program value falls below the ACAT I (or II) threshold. Risk, of which program cost is an important element, should be the dominant factor in ACAT decisions. Birkler et al. reach a similar conclusion in RAND research conducted for the Office of the Secretary of Defense and provide a more detailed breakdown of the specific elements of risk that should be considered.[13]

Programs Require Adequate Resources from the Beginning

A widely used aphorism in the acquisition world is that "programs that are born healthy stay healthy," or, said differently, programs require adequate funding and human capital from the beginning in order to succeed. Expediency often drives the services to other solutions, however. Sometimes programs begin with small budgets with the expectation that later, as additional funding becomes available, their budgets will be increased. They may also begin without the full complement of expertise necessary for success, with the expectation that talent can be added as it becomes available. The consensus among our experts emphasized the need to avoid these shortcuts and ensure that programs

---

[13] See Birkler et al., 2010.

are appropriately resourced from the outset. While we recognize the significant challenges to providing these resources, this is nonetheless particularly important for the success of large, complex programs, in particular.

More Fully Integrated Test and Evaluation

Although the test and evaluation community prides itself on rigorous, independent testing, a number of workshop participants noted that, in practice, its independence sometimes led to what amounts to new performance criteria for vehicles that were not represented in their ORDs, CDDs, or other requirements documents. Rather, the test and evaluation community uses its own criteria, according to these statements. While we are not aware of this affecting the vehicles examined in this study, experts assert that this late appearance of new performance criteria sometimes led to delays in final certification for vehicles and more often added to program cost and schedule delays as programs tried to satisfy the newly evolved standards. While acquisition personnel tended to see these criteria as "after-the-fact" requirements, the test and evaluation community viewed them as essential to the service's interests. Both viewpoints seem to have merit.

Service experts and officials in the Office of the Director, Defense Research and Engineering, agreed that testing and evaluation activities that were more closely integrated into the program's schedule could preserve the independence of the test and evaluation community while reducing the likelihood of new criteria emerging late in a program and imposing additional cost and schedule delays.

Observations Concerning Intelligence Support

The Office of the Under Secretary of Defense for Intelligence and the service intelligence organizations provide threat support to vehicle development programs (and to all materiel development efforts). The intelligence materials we examined over the course of this research employed a weapon-centric approach that identified the most likely weapons (e.g., those most often in the hands of likely adversaries) and the most stressful weapons (those that present the greatest danger to the U.S. systems under discussion, no matter how rare the enemy weapon might be).

These assessments could be made more valuable if they contained additional information that is likely already available within the Intelligence Community. For example, it would be useful for combat developers and the research, development, and acquisition community to have current information on common TTPs that U.S. forces might encounter on future battlefields so that they could appreciate the circumstances in which the enemy may employ its weapons against U.S. forces (potentially aiding in the design of the defensive equipment itself or off-vehicle countermeasures and defensive systems). In particular, the experiences of other countries would help the United States better anticipate particularly stressful enemy weapons and employment practices. For example, Russian Federation forces encountered large, sophisticated IEDs in Chechnya in the 1990s, yet U.S. forces were caught off guard when they encountered similar weapons after the invasion of Iraq in 2003. While it is likely that the Intelligence Community was aware of these weapons and how they were used in Chechnya, the requirements system either was not aware of them and the TTPs used in employing them or it discounted their importance in developing and updating requirements for weapon systems.

Trends

Equipping the armed services with tactical wheeled and ground combat vehicles will remain a challenging endeavor for the multitude of reasons cited throughout this monograph. Some developments are clearly positive and should help to ensure the acquisition of vehicles suitable for the anticipated circumstances. Some developments are clearly negative and complicate the task the services face in equipping their forces. Still other factors are ambiguous at this point but could prove to be positive or negative as their effects become more visible.

Positive Trends

The prevailing preference among program managers for relatively mature technologies at the beginning of the TD phase of vehicle programs is clearly positive. The practice reduces dependencies on immature technologies that can lead to cost growth and schedule slippage when they do not develop as quickly as estimated. The practice also increases the probability that those technologies that are central to the vehicle's success will be more fully developed than otherwise would be the case by the time the program faces EMD.

The services' increased appreciation of systems engineering expertise is another positive development. Both the Army and Marine Corps seem to recognize the centrality of systems engineering to program success and appear to be trying to grow their capacity in this field. The Army has almost certainly been chastened by its experience with FCS, when it sought to buy systems engineering and integration capabilities in the form of a lead systems integrator and found that it struggled to provide the governmental oversight for this function.

The services have renewed their efforts at better management practices and risk management, typified by knowledge points, competitive prototyping, gate reviews, portfolio reviews, requirement-stabilization initiatives, and other efforts described in Chapters Two and Three. To be sure, there is room for improvement here, perhaps with the adoption of a set of best practices at the program level or by insisting that all programs satisfy the criteria for being "born healthy," as described in Chapter Three.

Another positive sign lies in the responsiveness of the research, development, and acquisition communities. Collectively, they have shown a relatively recent improved ability to produce needed vehicles in a hurry; MRAP and M-ATV are good examples. They have also demonstrated responsiveness to Joint Urgent Operational Needs Statements and operational needs statements, fielding B-kit armor for HEMTTs and HMMWVs, among other responses.

M&S also holds great potential for improving vehicle designs, especially if this field evolves along the lines suggested in Chapter Five.

Negative Trends

If necessity continues to drive tactical wheeled vehicle requirements closer to those of their ground combat vehicle cousins, that will surely have the salutary effect of affording their crews greater protection and situational awareness, but these positive trends will be accompanied by complexity and cost growth. Given the experience of the past nine years of war, it seems likely that design criteria for ground combat vehicles will continue to influence the design of the tactical wheeled vehicle fleet. As a result, new vehicles will almost certainly be significantly more expensive than the ones they replace.

This phenomenon will probably manifest across all vehicle fleets as recapitalization and replacement go forward. In addition, as the case of the JLTV suggests, there may be a divergence in requirements between the Marine Corps and the Army to meet performance criteria exclusive to each service. If this occurs, the unit costs of the vehicles in question will probably increase, because of the loss of economies of scale when each service procures its own designs.

Then there is the persistent vulnerability of the vehicle fleets. As the GCV example suggests, this state of affairs can emphasize design criteria in favor of protection and compromise all other performance dimensions in the process. Technology-based solutions to mitigate vulnerability are expensive, whereas the enemy's countermeasures are relatively cheap. It has become nearly impossible to protect the vehicle fleets solely with onboard armor, situational awareness, and APS. At the same time, incorporating off-vehicle assets in trade-offs and calculations of vehicle requirements necessitates further assistance from the M&S community.

Uncertain Trends

The potential of robotics and autonomous systems, on its face, seems significant. Perhaps it is, but until the services advance these technologies and develop concepts for their application in roles that would reduce the threat to tactical wheeled and ground combat vehicles, their future utility remains in doubt. Removing soldiers and marines from harm's way is an important but perhaps insufficient contribution, especially if the costs associated with the systems in question rival or exceed those of the manned vehicles they replace.

The effects of the network on vehicles are another question mark. FCS revealed some current limitations. The key question is whether on- and off-vehicle capabilities can be integrated so that communication, situational awareness, protection, and power-generation requirements can be reduced for the vehicle fleets. If the answer is positive, then weight reductions, mobility improvements, fuel efficiency, and reduced unit costs might result. If the answer is negative, then the network would become another source for additional vehicle requirements, increasing complexity and cost in the process.

What Congress Can Do

In this monograph, we presented a number of strategic, technical, and business practice and process considerations that affect DoD's ability to field combat and tactical wheeled vehicle fleets that meet the country's needs. Some of these considerations take the form of things that Congress should pay attention to or do, whereas others frame and in some cases constrain DoD's ability to field these vehicle fleets. In this section, we provide a brief summary of these considerations and suggestions for how Congress might use them in its efforts to exercise its oversight responsibilities.

Making Strategic Choices

One strategic observation that goes beyond, but frames, the DoD research, development, and acquisition communities' immediate tasks is that it is very difficult to predict future threats over the expected life spans of vehicles now in production. DoD must make choices and accept risks due to the impossibility of designing vehicles that are optimal for all future threats. Congress should consider the implications of this fact as it interacts with DoD leaders on the development of vehicle fleets. This observation implies that the U.S. government will need to decide whether it wants to optimize vehicles for the most dangerous threat, the most likely threat, or based on some other criteria. During the Cold War, this was a simple choice; today, it is not. One alternative approach would be to develop vehicles that are adequate to counter several threats but not optimal for any. DoD leadership should clearly articulate what rationale it is using in vehicle fleet development (e.g., optimizing vehicles against a specific threat, as in the Cold War, or creating vehicles that are adequate for a spectrum of threats). Given the joint nature of conflict, this rationale should be considered by, if not standard across, each armed service. Congress should consider requiring that DoD present the strategic rationale for these choices *fleet wide*, as well as how each individual proposed vehicle fits within this rationale, rather than focusing on individual vehicle capabilities in the context of a narrowly defined operational requirement.

Ensuring That Key Technological Challenges Are Addressed

We highlighted four classes of technical challenges that currently affect, and for the foreseeable future will continue to affect, the ability of the defense research, development, and acquisition communities to field cutting-edge vehicles that meet the operational requirements of fielded forces: the need for improved protection, power generation, and fuel consumption, as well as the effect that sensors and networking have on the complexity of modern vehicles and the challenges that come with it. Because these are classes of problems that affect almost every vehicle (and many other systems) that DoD fields, they should be considered as such by Congress. This means that individual systems should have plans for each but, importantly, that attention is needed to them outside of individual programs of record as well. In particular, in its oversight role, Congress should ensure that defense programs to address each of these challenges are adequate. This would include, but not be limited to, working with DoD to ensure that the following areas are thoroughly addressed across the board:

- *Protection:* Improve ballistic protection (e.g., material sciences), advance APSs (e.g., requires material as well as nonmaterial approaches), and develop technical and tactical approaches to using the array of systems on the battlefield to improve the ability to identify, avoid, and defuse threats.

- *Power generation:* Improve both the amount of power that vehicles can provide (inherently and with ancillary power devices) and the efficiency of those items that require power in order to improve the ratio of power supply to demand. This is a complex task; the programs that develop and update items that require power do not fall under the purview of the programs that build the platforms into which they must be integrated. Improvements in this area would have far-reaching implications not only for the platforms in question but also for the number and types of equipment needed in the force (it would decrease them), the size and complexity of logistics operations, and cost.
- *Fuel and fuel efficiency:* The elements of this challenge include the types of fuels that tactical wheeled and ground combat vehicles can use, how efficient the engines are, and what can be done to improve performance in both categories. The bottom line is that improvements could have significant positive effects on both operations and costs.
- *Sensors, networking, and complexity:* Sensors and networking, when they work well, significantly enhance the capabilities of vehicles, but they also have the potential to create unanticipated technical challenges, as well as cost and schedule overruns. Addressing the problems that stem from complex systems is not trivial and in some cases not fully understood. However, approaches that include robust systems engineering capabilities and basic research, in some cases, are the best bets for anticipating problems and solving them early. Congress can play a positive role by insisting that good systems engineering plans be put in place for complex systems in its oversight of the DoD acquisition workforce personnel system, ensuring that the systems engineering workforce is adequate and properly used and that there is an examination of research and development portfolios to ensure that critical areas affecting complexity are identified and efforts funded.

Congress should consider making all four of these areas focal points of its interactions with DoD on research and development, new systems, and modifications to existing systems.

Providing Guidance and Assistance on Research, Development, Acquisition, Testing and Evaluation, and Business Practices and Policies

This monograph identified seven areas in which business practices, processes, and policy changes could significantly enhance the research, development, acquisition, and test and evaluation communities' ability to better use resources and time to accomplish their tasks. Some of these challenges can be addressed—and may be in the process of being addressed—by DoD (e.g., how cost estimation is done, how programs are staffed and supported for success, how testing and evaluation are done). Some may require congressional action in the form of guidance, changes to laws, or clarification of congressional intent with a focus on regulations (e.g., adopting ACAT decision practices that more realistically address risk rather than using cost as a proxy for risk).[14] And some if not all have cost implications that Congress should factor into the way it oversees vehicle fleet development (e.g., the rising costs of tactical wheeled vehicles). In all seven cases, Congress may decide that the changes required for progress demand that it play some role. Furthermore, in all seven cases, Congress should consider asking for updates and challenging DoD to make or recommend changes.

The Importance of Improved Modeling and Simulation

Finally, a comprehensive M&S capability and leaders who understand how to use it will be an essential tool in everything from establishing future requirements to research and development to engineering and program design and manufacturing. DoD and the services should consider improvements to their already substantial capabilities along the lines outlined here, which will require support and guidance from Congress.

---

[14] Birkler et al., 2010.

Bibliography

The findings reported in this monograph were supported by a range of documents, both publicly available and not. Although the research effort was far-reaching and the content of these documents (listed in this bibliography) helped develop and reinforce the conclusions, the monograph uses unrestricted material only.

Acquisition Community Connection, Defense Acquisition University, "Joint Capabilities Integration and Development System (JCIDS)," last updated August 3, 2009. As of January 4, 2011:
https://acc.dau.mil/CommunityBrowser.aspx?id=164563&lang=en-US

Adams, Thomas K., *The Army After Next: The First Postindustrial Army*, Stanford, Calif.: Stanford University Press, 2008.

Anulare, Louis, Program Executive Office, Combat Support and Combat Service Support, "NDAA Section 222," Washington, D.C., briefing, May 12, 2010.

Army Capabilities Integration Center, "Capabilities Assessment and RAM Division (CARD) Information Brief," undated.

———, "Combat Developer Reliability, Availability & Maintainability for the Paladin Integrated Management (PIM) Family of Vehicles (FOV) Certification and Summary Sheet," February 4, 2008

———, "TRADOC Strategic Framework Slide Writer's Guide Final Version 1.1," September 10, 2008.

———, *TRADOC Capabilities-Based Assessment (CBA) Guide Version 3.1*, May 10, 2010a.

———, "TRADOC Capabilities Production Document (CPD) Writer's Guide Version 1.9," July 1, 2010b.

Army Capabilities Integration Center, Research, Development, and Engineering Command, and Deputy Chief of Staff G-4, *Power and Energy Strategy White Paper*, April 1, 2010. As of December 28, 2010:
http://www.arcic.army.mil/Docs/PE%20Strategy%20Apr%202010.pdf

Army Force Management School, "Force Management Update," *AFMS News Letter*, July 2010.

Assistant Secretary of the Army for Acquisition, Logistics, and Technology, *2007–2008 Weapon Systems, ADA473205*, Washington, D.C., 2007.

Birkler, John, Mark V. Arena, Irv Blickstein, Jeffrey A. Drezner, Susan M. Gates, Melinda Huang, Robert Murphy, Charles Nemfakos, and Susan K. Woodward, *From Marginal Adjustments to Meaningful Change: Rethinking Weapon System Acquisition*, Santa Monica, Calif.: RAND Corporation, MG-1020-OSD, 2010. As of December 28, 2010: http://www.rand.org/pubs/monographs/MG1020.html

Brannen, Kate, "DoD Adds JLTV Subvariant: More Protection, Same Weight," *DefenseNews*, June 7, 2010a.

———, "Mobility vs. Survivability: JLTV Could Suffer as U.S. Army, Marines Diverge," *DefenseNews*, June 7, 2010b.

———, "U.S. Army's GCV Delay: Lesson Unlearned," *DefenseNews*, August 27, 2010c.

———, "Efficiency Push Could Threaten U.S. JLTV," *DefenseNews*, October 7, 2010d.

———, "Gen. Peter Chiarelli: Vice Chief of Staff, U.S. Army," *DefenseNews*, October 25, 2010e.

———, "U.S. Army Reaffirms JLTV Commitment," *DefenseNews*, October 26, 2010f.

———, "'Chimney' Deflects IEDs: Humvees Survive 3 Aberdeen Tests," *DefenseNews*, November 1, 2010g.

Capability Production Document (CPD) for the M109 Family of Vehicles (FOV) Increment 1 ACAT II, August 2009.

Carter, Ashton B., Under Secretary of Defense for Acquisition, Technology, and Logistics, "Ground Combat Vehicle (GCV) Material Development Decision (MDD) Acquisition Decision Memorandum," memorandum to the Secretary of the Army, May 11, 2010a, not available to the general public.

———, *Army Modernization Follow-On to Future Combat Systems (FCS)*, Washington, D.C., May 14, 2010b.

———, "Better Buying Power: Guidance for Obtaining Greater Efficiency and Productivity in Defense Spending," memorandum to acquisition professionals, September 14, 2010c.

Catto, Maj. Gen. William, U.S. Marine Corps, testimony before the House Committee on Armed Services, Subcommittee on Tactical Air and Land Forces, on Marine Corps force protection efforts, June 15, 2006.

Chairman of the Joint Chiefs of Staff Instruction 3170.01G, Joint Capabilities Integration and Development System, March 1, 2009.

Chassé, Corey B., "PIM: The Next Generation Paladin," *Fires Bulletin*, January–February 2008.

Commanding General, Marine Corps Combat Development, "Operational Requirements Document (ORD) for the Medium Tactical Vehicle Replacement (MTVR) (No. MOB 211.4.2), Change 4," Quantico, Va., March 12, 1998.

Cox, Matthew, "Casey: Make Ground Combat Vehicle Lighter," *Army Times*, June 21, 2010. As of December 28, 2010:
http://www.armytimes.com/news/2010/06/army_gcv_casey_062110w/

Defense Acquisition University, "Data Item Description (DID) System Engineering Management Plan (SEMP)," web page, October 11, 2002. As of December 28, 2010:
https://acc.dau.mil/CommunityBrowser.aspx?id=59153

―――, "Technology Insertion," web page, May 20, 2003. As of December 28, 2010:
https://acc.dau.mil/CommunityBrowser.aspx?id=32703

―――, "Knowledge-Based Acquisition," web page, June 21, 2004. As of December 28, 2010:
https://acc.dau.mil/CommunityBrowser.aspx?id=24660

―――, "Systems Engineering Plan (SEP)," *ACQuipedia*, April 19, 2005. As of December 28, 2010:
https://acc.dau.mil/CommunityBrowser.aspx?id=29041

―――, "Naval Gate Review Process," briefing, March 19, 2009. As of December 28, 2010:
https://acc.dau.mil/CommunityBrowser.aspx?id=272437

DiPetto, Chris, Office of the Deputy Under Secretary of Defense for Acquisition and Technology, testimony before the House Committee on Armed Services, Subcommittee on Readiness, March 13, 2008a. As of December 28, 2010:
http://www.dod.gov/dodgc/olc/docs/testDipetto080313.pdf

―――, "DoD Energy Demand: Addressing the Unintended Consequences," September 2008b.

Dreyer, Paul, and Paul K. Davis, *A Portfolio-Analysis Tool for Missile Defense: Methodology and User's Manual*, Santa Monica, Calif.: RAND Corporation, TR-262-MDA, 2005. As of December 28, 2010:
http://www.rand.org/pubs/technical_reports/TR262.html

―――, *RAND's Portfolio Analysis Tool (PAT): Theory, Methods, and Reference Manual*, Santa Monica, Calif.: RAND Corporation, TR-756-OSD, 2009. As of December 28, 2010:
http://www.rand.org/pubs/technical_reports/TR756.html

"DRS and Allison Transmission Announce Strategic Partnership," *ASD News*, November 2, 2010. As of December 28, 2010:
http://www.asdnews.com/news/31560/DRS_and_Allison_Transmission_Announce_Strategic_Partnership.htm

DRS Technologies, "DRS Technologies Provides Critical Power Technology to United States Marine Corps for On-the-Move Power Needs of Tactical-Wheeled Vehicles," news release, June 28, 2010a. As of December 28, 2010:
http://news.drs.com/news/drstechnologies+unitedstatesmarinecorps+onthemovepower+tacticalwheeledvehicles.htm

————, "HMMWV Power Now, Anywhere!" Huntsville, Ala., August 19, 2010b. As of December 28, 2010:
http://www.drs.com/Products/TS/PDF/obvphmmvv.pdf

Duma, David W., Principal Deputy Director, Operational Test and Evaluation, Office of the Secretary of Defense, testimony before the Senate Committee on Armed Services, Subcommittee on Airland, April 15, 2010.

Fastabend, BG David, Deputy Director, U.S. Army Training and Doctrine Command Futures Center, "The Army in Joint Operations: The Army's Future Force Capstone Concept 2015–2024," video, July 25, 2005.

Feickert, Andrew, *The Marines' Expeditionary Fighting Vehicle (EFV): Background and Issues for Congress*, Washington, D.C.: Congressional Research Service, RS22947, August 3, 2009.

————, *Joint Light Tactical Vehicle (JLTV): Background and Issues for Congress*, Washington, D.C.: Congressional Research Service, RS22942, September 17, 2010.

Francis, Paul L., Director of Acquisition and Sourcing Management, U.S. General Accounting Office, *Defense Acquisitions: The Army's Future Combat Systems Features, Risks, and Alternatives*, testimony before the House Committee on Armed Services, Subcommittee on Tactical Air and Land Forces, Washington, D.C.: U.S. General Accounting Office, GAO-04-635T, April 1, 2004.

GAO—*see* U.S. Government Accountability Office.

General Dynamics Amphibious Systems, Diminishing Manufacturing and Material Shortages Integrated Product Team, *Expeditionary Fighting Vehicle Diminishing Manufacturing Sources and Material Shortages (DMSMS) Management Plan*, Woodbridge, Va., CDRL S066, Revision 05-00, June 1, 2005.

General Dynamics Land Systems, *Expeditionary Fighting Vehicle*, draft, April 14, 2010.

Glenn, Russell W., Jody Jacobs, Brian Nichiporuk, Christopher Paul, Barbara Raymond, Randall Steeb, and Harry J. Thie, *Preparing for the Proven Inevitable: An Urban Operations Training Strategy for America's Joint Force*, Santa Monica, Calif.: RAND Corporation, MG-439-OSD/JFCOM, 2006. As of December 28, 2010:
http://www.rand.org/pubs/monographs/MG439.html

Gonzales, Jose M., Acting Deputy Director, Land Warfare and Munitions, "Joint Light Tactical Vehicle (JLTV) Overarching Integrated Product Team (OIPT) Report," memorandum for the Under Secretary of Defense for Acquisition, Technology, and Logistics, May 14, 2010.

Goodman, Glenn W., Jr., "Joint Light Tactical Vehicle," web page, undated. As of January 7, 2010:
http://www.marcorsyscom.usmc.mil/peolandsystems/jltv.aspx

———, "S&T and Cost Estimating," in Scott R. Gourley, *Nearly Four Years in Operation: Program Executive Officer Land Systems Marine Corps Looks Ahead to the Future*, Quantico, Va.: Program Executive Office, Land Systems, U.S. Marine Corps, 2010, p. 5. As of January 7, 2010:
http://www.marcorsyscom.usmc.mil/peolandsystems/docs/PEO%20LS%20 Overview.pdf

Gourley, Scott R., "The Future of Tactical Wheeled Vehicles," Washington, D.C.: Headquarters, U.S. Department of the Army, June 1, 2010.

Greenberg, Marc, and James Gates, "Analysis of Alternatives," Washington, D.C.: National Defense University, April 2006.

Greig, Scot, Matt Donohue, Kristopher Gardner, Thomas D'Agostino, and Roger Pratt, *Future Combat Systems (FCS): Technology Readiness Assessment (TRA) Executive Summary*, Washington, D.C.: Office of the Deputy Assistant Secretary of the Army for Research and Technology, predecisional draft, May 2009, not available to the general public.

Gutting, Glenn E., "Inside Urban Resolve 2015 and Omni Fusion 2006," TRADOC News Service, October 25, 2006. As of December 28, 2010:
http://www.tradoc.army.mil/pao/Web_specials/ur2015/2015_omnifu06.htm

"HASC Leaders Write Gates About Army Abrams, Bradley and Stryker Funding," *Defense News Daily*, October 1, 2008.

Headquarters, U.S. Department of Army, "Annex B 155mm Howitzer Improvement Program (HIP) Operational Mode Summary/Mission Profile," undated.

———, "RAM Rationale Report (RRR) for the Howitzer Improvement Program (HIP)," February 24, 1988a.

———, "Required Operational Capability (ROC) for the M109 Howitzer Improvement Program (HIP)," August 2, 1988b.

————, "Heavy Expanded Mobility Tactical Truck (HEMTT) Operational Requirements Document (ORD)," Fort Monroe, Va., August 27, 1998.

————, *Simulation Support Planning and Plans*, Pamphlet 5-12, March 2, 2005.

————, "Heavy Expanded Mobility Tactical Truck (HEMTT) Critical Operational Issues and Criteria (COIC) Version 8," Fort Monroe, Va., September 18, 2006.

————, *Operations*, Field Manual 3-0, Washington, D.C., February 27, 2008.

————, "Approval of the Capability Production Department (CPD) for the M109 Family of Vehicles (FoV)," August 6, 2009a.

————, *Final Report: Functional Solution Analysis (FSA) for Medium and Heavy Tactical Wheeled Vehicles (TWV) and Trailers*, September 25, 2009b.

————, "Critical Operational Issues and Criteria for the M109 Family of Vehicles," November 11, 2009c.

————, "Approval of Functional Solution Analysis (FSA) for Medium and Heavy Tactical Wheeled Vehicles (TWV) and Trailers," December 18, 2009d.

————, "Stryker Family of Vehicles," briefing slides, May 10, 2010, not available to the general public.

————, "Brigade Combat Team (BCT) Modernization," memorandum to the Under Secretary of Defense for Acquisition, Technology and Logistics, May 12, 2010, not available to the general public.

————, *Army Truck Program (Tactical Wheeled Vehicle Acquisition Strategy) Report to the Congress*, June 2010a.

————, "FY11 Warfighter Outcomes (WFO) Analysis Results," July 1, 2010b.

————, "Capability Production Document (CPD) for Heavy Expanded Mobility Tactical Truck (HEMTT)," August 5, 2010c, not available to the general public.

————, *NDAA Section 222 Modernization of Ground Combat and Armored Tactical Vehicle Study*, October 20, 2010d, not available to the general public.

Headquarters, U.S. Department of the Army, Army Requirements Oversight Council, "JLTV Family of Vehicles," briefing, March 28, 2007.

Headquarters, U.S. Department of the Army, Deputy Chief of Staff for Programs, G-8, *HMMWV Requirements & O/H Data*, undated(a).

————, "JLTV Within Current HMMWV Funding," briefing slide, undated(b).

————, *The Army Tactical Wheeled Vehicle Investment Strategy*, white paper, October 30, 2009.

————, *2010 Army Modernization Strategy*, April 23, 2010. As of December 28, 2010:
http://www.bctmod.army.mil/program/AMS2010_lq.pdf

Headquarters, U.S. Department of the Army, Office of the Deputy Chief of Staff for Plans and Operations, Force Development, *Abrams Modernization Program M1A2, Operational Requirements Document*, Washington, D.C., January 1994.

Headquarters, U.S. Department of the Army, U.S. Department of the Navy, and U.S. Marine Corps, "JLTV/MRAP Comparison," briefing slides, undated.

————, *Joint Light Tactical Vehicle (JLTV) Industry Day*, August 2, 2006.

"Highlights of the FY2009 NDAA," *Federal Contracts Report*, Vol. 89, No. 22, June 10, 2008. As of December 28, 2010:
http://www.ndia.org/Advocacy/Resources/Pages/Legislative_Alerts.aspx

"HMMWV (High Mobility Multipurpose Wheeled Vehicle)," *U.S. Army Factfiles*, undated. As of December 28, 2010:
http://www.army.mil/factfiles/equipment/wheeled/hmmwv.html

Holm, David, "Milestone A: Costing Joint Light Tactical Vehicle Program," briefing slides, undated.

Itoh, Masato, "New EFV Prototype Tests at Camp Pendleton," *Ground Combat Technology*, Vol. 1, No. 2, August 2010. As of December 28, 2010:
http://www.military-information-technology.com/gct-home/264-gct-2010-volume-1-issue-2-august/3287-new-efv-prototype-tests-at-camp-pendleton-.html

John, Ashley, "PEO CS&CSS JLTV Program Receives Top 5 DoD Program Award," *Accelerate*, Summer 2010, pp. 69–71. As of December 28, 2010:
http://tardec.army.mil/Documents/TARDEC_0910_accelerate_Summer_2010.pdf

Johnson, LTG Albert, "Test and Evaluation Master Plan (TEMP) Addendum for the Heavy Expanded Mobility Tactical Truck HEMTT A4 Long Term Armoring Strategy (LTAS)," April 2, 2007.

Johnson, LTC Allen, "PM Heavy Tactical Vehicle," briefing presented at the 2010 Tactical Wheeled Vehicles Conference, Monterey, Calif., February 7–9, 2010. As of December 28, 2010:
http://www.dtic.mil/ndia/2010tactical/PMfeb9AllenJohnson_HTV.pdf

Keeter, H., "Marine Corps' AAAV to Be Renamed 'Expeditionary Fighting Vehicle' (Advanced Amphibious Assault Vehicle)," *Defense Daily*, July 1, 2003.

Kimball, Linda, "Modeling and Simulation Tools to Support the Challenges of T&E in an Urban Environment," briefing, U.S. Army Materiel Systems Analysis Activity, Huntsville, Ala., June 15, 2010.

Koerner, Heather, and Gordon McDonald, *A Conceptual Framework for the U.S. Army Tactical Wheeled Vehicle Optimization Model*, thesis, Monterey, Calif.: Naval Postgraduate School, June 2007.

Lamb, Todd, Program Manager, Development, "Stryker Modernization Update (AUSA)," briefing presented at the meeting of the Association of the United States Army, October 6, 2009.

Lennox, LTG Robert P., LTG William N. Phillips, and David M. Markowitz, U.S. Army, "On Army Acquisition and Modernization Programs," testimony before the House Committee on Armed Services, Subcommittee on Air and Land Forces, March 10, 2010.

Lopez, C. Todd, "M-ATVs to Replace Humvees in Afghanistan, Vice Says," Army News Service, April 15, 2010a. As of December 28, 2010:
http://www.army.mil/-news/2010/04/15/
37367-m-atvs-to-replace-humvees-in-afghanistan-vice-says/index.html

———, "GCV Must Be Safe, Affordable, Full-Spectrum Capable," Army News Service, October 4, 2010b. As of December 28, 2010:
http://www.army.mil/-news/2010/10/04/
46074-gcv-must-be-safe-affordable-full-spectrum-capable/

———, "Saving Energy Saves Lives Says New Army Exec," Army News Service, October 13, 2010c. As of December 28, 2010:
http://www.army.mil/-news/2010/10/13/
46504-saving-energy-saves-lives-says-new-army-exec/

"M1 Abrams Main Battle Tank," MilitaryPeriscope.com, last updated May 1, 2009. As of December 28, 2010:
http://www.militaryperiscope.com/weapons/gcv/tanks/w0003593.html

Martell, Mike, Heavy Brigade Combat Team, DPM Abrams, "Abrams Modernization RAND Study," briefing, May 12, 2010, not available to the general public.

Matsumura, John, Randall Steeb, John Gordon IV, Thomas J. Herbert, Russell W. Glenn, and Paul Steinberg, *Lightning Over Water: Sharpening America's Light Forces for Rapid Reaction Missions*, Santa Monica, Calif.: RAND Corporation, MR-1196-A/OSD, 2000. As of December 28, 2010:
http://www.rand.org/pubs/monograph_reports/MR1196.html

Matsumura, John, Randall Steeb, Thomas J. Herbert, John Gordon IV, Carl Rhodes, Russell W. Glenn, Michael Barbero, Frederick J. Gellert, Phyllis Kantar, Gail Halverson, Robert Cochran, and Paul Steinberg, *Exploring Advanced Technologies for the Future Combat Systems Program*, Santa Monica, Calif.: RAND Corporation, MR-1332-A, 2002. As of December 28, 2010:
http://www.rand.org/pubs/monograph_reports/MR1332.html

Meyerle, Jerry, and Carter Malkasian, *Insurgent Tactics in Southern Afghanistan 2005–2008*, CNA Strategic Studies Division, CRM D0020729.A2/Final, August 2009, not available to the general public.

Moore, Louis, "Evolving the Army's Strategy for the Light Tactical Vehicle (LTV) Fleet," Santa Monica, Calif.: RAND Corporation, undated.

Moore, Louis, Elvira Loredo, and Keenan Yoho, "Evolving the Army's Strategy for the Light Tactical Vehicle (LTV) Fleet," Santa Monica, Calif.: RAND Corporation, briefing, June 19, 2007, not available to the general public.

Myers, COL John S., Project Manager, Joint Combined Support Systems, "Joint Light Tactical Vehicle (JLTV)," briefing slides, October 2006a.

———, "Advanced Planning Brief to Industry," briefing, October 27, 2006b.

———, "Joint Combat Support Systems," briefing at the National Defense Industrial Association Tactical Wheeled Vehicles Conference, Monterey, Calif., February 9, 2010. As of December 28, 2010: http://www.dtic.mil/ndia/2010tactical/feb9JohnMyersTWVNDIAConf.pdf

Myers, Steve, and Ben Garza, "TWV Transformational Efforts," briefing at the National Defense Industrial Association Tactical Wheeled Vehicles Conference, February 6, 2007.

North Atlantic Treaty Organization, *NATO Reference Mobility Model*, Brussels, RTO-TR-AVT-107, 2002.

Office of the Assistant Secretary of Defense for Public Affairs, "Future Combat System [FCS] Program Transitions to Army Brigade Combat Team Modernization," news release, No. 451-09, June 23, 2009. As of December 28, 2010: http://www.defense.gov/Releases/Release.aspx?ReleaseID=12763

Office of Naval Research, "O&S/LCC Cost Reductions Through Technology Insertions," briefing, January 4, 2011.

Office of the Secretary of Defense, *Defense Acquisition Transformation: Report to Congress, John Warner National Defense Authorization Act, Fiscal Year 2007, Section 804*, July 2007. As of December 28, 2010: http://www.defense.gov/pubs/pdfs/804JulFinalReport_to_Congress.pdf

Osborn, Kris, "Army Leaders Brief Industry on Ground Combat Vehicles," Army News Service, October 5, 2010. As of December 28, 2010: http://www.army.mil/-news/2010/10/05/46163-army-leaders-brief-industry-on-ground-combat-vehicle

Oshkosh Corporation, "Victory Is Dependent on Supplying Troops in the Field," brochure, May 2005.

Peltz, Eric, John Halliday, and Aimee Bower, *Speed and Power: Toward an Expeditionary Army*, Santa Monica, Calif.: RAND Corporation, MR-1755-A, 2003. As of December 28, 2010: http://www.rand.org/pubs/monograph_reports/MR1755.html

Perry, Walter L., Bruce R. Pirnie, and John Gordon IV, *Issues Raised During the 1998 Army After Next Spring Wargame*, Santa Monica, Calif.: RAND Corporation, MR-1023-A, 1999. As of December 28, 2010: http://www.rand.org/pubs/monograph_reports/MR1023.html

Petermann, Wolfgang, and Ruben Garza, "JLTV Information Briefing to Industry," briefing at pre-proposal conference, February 19–21, 2008.

Product Manager, Fire Support Platforms, "PIM Technology Insertion Strategy," briefing slides, undated.

Program Manager, Stryker Brigade Combat Team, "Stryker Brigade Combat Team (SBCT) Modernization Strategy," briefing, September 22, 2010.

Public Law 111-84, National Defense Authorization Act for Fiscal Year 2010, October 28, 2009.

Reuter, Matthew B., *Reliability Analysis and Modeling of the U.S. Marine Corps Medium Tactical Wheeled Vehicle in Operation Iraqi Freedom*, thesis, Monterey, Calif.: Naval Post Graduate School, September 2007.

Research and Innovative Technology Administration, Bureau of Transportation Statistics, "Number of U.S. Aircraft, Vehicles, Vessels, and Other Conveyances," 2010. As of December 28, 2010:
http://www.bts.gov/publications/national_transportation_statistics/html/table_01_11.html

Ricks, Thomas E., "Inside an Afghan Battle Gone Wrong: What Happened at Wanat?" *Foreign Policy*, January 28, 2009.

Rogers, Paul, Executive Director of Research, U.S. Army Tank Automotive Research, Development and Engineering Center, Research, Development, and Engineering Command, "FCS Technology Insertion and Transition," briefing presented at the eighth annual Science and Engineering Technology Conference, North Charleston, S.C., April 18, 2007. As of December 28, 2010:
http://www.dtic.mil/cgi-bin/GetTRDoc?Location=U2&doc=GetTRDoc.pdf&AD=ADA469061

———, "TARDEC's Ground Vehicle Power and Energy Overview," briefing presented at the Michigan Defense and Innovation Symposium, Livonia, Mich., November 17, 2008. As of November 4, 2010:
http://www.dtic.mil/cgi-bin/GetTRDoc?Location=U2&doc=GetTRDoc.pdf&AD=ADA504866

Rowan, Jim, Deputy Commandant, U.S. Army Engineer School, "Update to Emeritus Leaders," briefing, September 29, 2010. As of December 28, 2010:
http://www.usace.army.mil/about/LeadersEmeritus/Documents/Read_Ahead/Regiment_USACE_Briefing.pdf

Scales, Robert H., "A Vehicle for Modern Times: What the Next Combat Vehicle Should Look Like," *Armed Forces Journal*, December 2009–January 2010.

Schumitz, COL Robert W., Stryker Brigade Combat Team Project Manager, "Stryker Double-V Hull (DVH) IIPT," briefing, June 17, 2010a.

————, "PM Stryker Brigade Combat Team," briefing for the Program Executive Office, Ground Combat Systems, advance planning brief for industry panel, October 22, 2010b. As of December 28, 2010:
http://contracting.tacom.army.mil/future_buys/FY10/Davis.pdf

Shalal-Esa, Andrea, "Army, Pentagon at Odds Over New Vehicle Program," Reuters, February 17, 2010. As of November 4, 2010:
http://www.reuters.com/article/idUSTRE61G5NT20100217

Shinseki, GEN Eric K., Chief of Staff, U.S. Army, statement on the status of forces before the Senate Committee on Armed Services, October 26, 1999.

————, "The Army Transformation: A Historic Opportunity," *2001–02 Army Green Book*, Arlington, Va.: Association of the United States Army, 2001.

Silberglitt, Richard, Lance Sherry, Carolyn Wong, Michael S. Tseng, Emile Ettedgui, Aaron Watts, and Geoffrey Strothard, *Portfolio Analysis and Management for Naval Research and Development*, Santa Monica, Calif.: RAND Corporation, MG-271-NAVY, 2004. As of December 28, 2010:
http://www.rand.org/pubs/monographs/MG271.html

Steeb, Randall, John Matsumura, Paul Steinberg, Thomas J. Herbert, Phyllis Kantar, and Patrick Bogue, *Examining the Army's Future Warrior: Force-on-Force Simulation of Candidate Technologies*, Santa Monica, Calif.: RAND Corporation, MG-140-A, 2004. As of December 28, 2010:
http://www.rand.org/pubs/monographs/MG140.html

Steeb, Randall, et al., *A Simulation-Based Exploration of Options to Protect Small Units, Based on a Case Study of the Battle of Wanat, Afghanistan*, Santa Monica, Calif.: RAND Corporation, 2010, not available to the general public.

Sullivan, Michael J., Director of Acquisition and Sourcing Management, U.S. Government Accountability Office, *Defense Acquisitions: Opportunities and Challenges for Army Ground Force Modernization Efforts*, testimony before the Senate Committee on Armed Services, Subcommittee on Airland, Washington, D.C.: U.S. Government Accountability Office, GAO-10-603T, April 15, 2010.

Tan, Michelle, "Action Taken in COP Keating Attack," *Army Times*, February 11, 2010.

Technical Joint Cross Service Group, *Analysis and Recommendations*, Vol. 12, May 19, 2005. As of December 28, 2010:
http://www.defense.gov/brac/pdf/12_techfinalreport5_20_05o.pdf

Tiron, Roxana, "$400 per Gallon Gas to Drive Debate Over Cost of War in Afghanistan," *The Hill*, October 15, 2009. As of December 28, 2010:
http://thehill.com/homenews/administration/
63407-400gallon-gas-another-cost-of-war-in-afghanistan-

U.S. Army, *Change-1 of Operational Requirements Document for the Family of Stryker Vehicles ACAT I*, November 2007, not available to the general public.

————, *System Training Plan for the M109 Family of Vehicles, Increment 1*, March 11, 2008.

————, *Critical Operational Issues and Criteria for the M109 Family of Vehicles (FOV) for the Test and Evaluation Master Plan Supporting Milestone C*, 2009a.

————, *RAM Analysis to Support Requirements for Paladin Howitzer CPD Revision 1*, March 4, 2009b, not available to the general public.

————, "Initial Capabilities Document (ICD) for Ground Combat Vehicle (GCV)," prepared for materiel development decision, draft, v1.3, September 29, 2009c, not available to the general public.

————, *Capability Development Document for Ground Combat Vehicle Core, Prepared for Milestone B Decision*, predecisional draft, v1.13, November 4, 2009d, not available to the general public.

————, "Annex C: Infantry Fighting Vehicle (IFV) Variant as of 06 Nov 09 ver 22," in *Capability Development Document for Ground Combat Vehicle Core, Prepared for Milestone B Decision*, predecisional draft, November 6, 2009e, not available to the general public.

————, "Joint Light Tactical Vehicle (JLTV)," briefing slides, 2010a.

————, "Tactical Wheeled Vehicle Portfolio Review to the Vice Chief of Staff of the Army," January 15, 2010b, not available to the general public.

————, *Initial Capabilities Document (ICD) for Ground Combat Vehicle (GCV)*, February 24, 2010c, not available to the general public.

————, *The Army Tactical Wheeled Vehicle Strategy*, October 2010d. As of January 3, 2011:
http://defensenews.com/blogs/ausa/files/2010/10/The_Army_TWV_Strategy_lq.pdf

————, *Capability Development Document (CDD) for the Abrams Tank Operational Requirements Document*, draft v1.6, October 14, 2010e, not available to the general public.

U.S. Army and U.S. Marine Corps, *Capability Development Document for Joint Light Tactical Vehicle (JLTV)*, v2.5, March 14, 2007, not available to the general public.

————, *Capability Development Document for Joint Light Tactical Vehicle (JLTV)*, v2.7a, November 15, 2007, not available to the general public.

U.S. Army Capabilities Integration Center, "FY11 Tier 1 Warfighter Outcomes, BIG Five Warfighter Outcomes," May 2010a, not available to the general public.

————, "OSD (DDR&E) and RAND Assessment of Activities for Technology Modernization of the Combat Vehicle and Armored Tactical Wheeled Fleets," briefing, October 20–21, 2010b.

U.S. Army Contracting Command, Warren, Mich., Solicitation Number W56HZV-11-R-0001, November 30, 2010. As of December 28, 2010: http://contracting.tacom.army.mil/confls/sol/W56HZV11R0001/0000.pdf

U.S. Army, Heavy Brigade Combat Team, and Product Manager, Fire Support Platforms, "PIM Technology Insertion Strategy," May 10, 2010, not available to the general public.

U.S. Army, Heavy Brigade Combat Team and Team Abrams, "Abrams Modernization Program Chronology," May 10, 2010, not available to the general public.

U.S. Army, Program Executive Office, Ground Combat Systems, "NDAA Section 222 DDR&E/RAND Study," May 12, 2010, not available to the general public.

U.S. Army, Program Executive Office, Integration, "Things You Need to Know About Ground Combat Vehicle Industry Day 2," welcome letter, Ground Combat Vehicle Industry Conference, November 2009a.

———, "Industry Day #2," Ground Combat Vehicle Conference agenda, November 23, 2009b, not available to the general public.

———, *B2 B3 Armor Materials*, Washington, D.C., PEO I Case 09-9161, November 16, 2009.

———, *Technology Development Strategy: Ground Combat Vehicle*, draft, April 9, 2010.

U.S. Army Research, Development, and Engineering Command and U.S. Army Tank Automotive Research, Development, and Engineering Center, "JLTV Briefing to Industry," May 2009.

U.S. Army Tank-Automotive and Armaments Command, "JLTV Purchase Description," briefing, undated. As of December 28, 2010: http://contracting.tacom.army.mil/majorsys/jltv/Day%201%20-%201500%20-%20PD%20Over%20-%20(various).ppt

———, *2006 Tactical Wheeled Vehicle Fleet*, February 1, 2006.

———, "Distribution A: Ground Combat Vehicle Industry Day," October 16, 2009.

———, *Draft Purchase Description (PD) for Joint Light Tactical Vehicle (JLTV) Family of Vehicles*, version 2.3, April 15, 2010. As of January 4, 2011: http://contracting.tacom.army.mil/majorsys/jltv/Day%201%20-%201500%20-%20PD%20Over%20-%20(various).ppt

U.S. Army Tank Automotive Research, Development, and Engineering Center, *RAND NDAA Assessment*, May 13, 2010a, not available to the general public.

————, "Overview: Tank Automotive Research, Development & Engineering Center," briefing presented at the Automotive Supplier Technology Forum, Troy, Mich., June 22, 2010b.

U.S. Army Training and Doctrine Command, *Initial Capabilities Document for Ground Combat Forces Light Tactical Mobility*, draft, v1.0, May 23, 2006, not available to the general public.

————, "Capabilities Development for Rapid Transition (CDRT)," in *2010 Army Posture Statement*, December 2009. As of September 31, 2010:
https://secureweb2.hqda.pentagon.mil/vdas_armyposturestatement/2010/
information_papers/Capabilities_Development_for_Rapid_Transition_
(CDRT).asp

U.S. Army Training and Doctrine Command Analysis Center, *Joint Light Tactical Vehicle (JLTV) Analysis of Alternatives (AoA) Study Plan Briefing*, September 17, 2009.

————, "Ground Combat Vehicle Analysis of Alternatives (GCV AoA) for Milestone A," memorandum to the director, Headquarters, U.S. Department of the Army, Deputy Chief of Staff, G3/5/7, Washington, D.C., April 22, 2010a.

————, "Ground Combat Vehicle (GCV) Analysis of Alternatives (AoA) Capability Gap Mitigation Analysis," June 17, 2010b, not available to the general public.

U.S. Army Training and Doctrine Command Futures Center, "Combat Developer Reliability, Availability and Maintainability Analysis for the Paladin Operations Center Vehicle (POCV) Certification Sheet," January 6, 2006.

U.S. Army Training and Doctrine Command Regulation 71-120, Concept Development, Experimentation, and Requirements Determination, October 6, 2009. As of December 28, 2010:
http://www.tradoc.army.mil/TPUBS/regs/tr71-20.pdf

U.S. Department of Defense, *Technology Readiness Levels in the Department of Defense (DoD), 2006 Defense Acquisition Guidebook*, Washington, D.C., 2006.

————, *Quadrennial Defense Review Report*, Washington, D.C., February 2010. As of December 28, 2010:
http://www.defense.gov/qdr/images/QDR_as_of_12Feb10_1000.pdf

U.S. Department of Defense, Joint Requirements Oversight Council, "Expeditionary Fighting Vehicle Capabilities Production Document," memorandum to the Under Secretary of Defense for Acquisition, Technology, and Logistics and Assistant Commandant of the Marine Corps, June 2, 2006.

————, "Expeditionary Fighting Vehicle Nunn-McCurdy Certification," memorandum to the Under Secretary of Defense for Acquisition, Technology, and Logistics and Assistant Commandant of the Marine Corps, May 8, 2007.

U.S. Department of Defense, Program Executive Office, Integration, *Ground Combat Vehicle White Paper*, Washington, D.C., draft, v11a, August 1, 2009, not available to the general public.

———, *Ground Combat Vehicle Operational Mode Summary/Mission Profile for Ground Combat Vehicle (GCV)*, Warren, Mich., draft, October 30, 2009, not available to the general public.

U.S. Department of Defense, U.S. Department of the Navy, and Headquarters U.S. Marine Corps, "Initial Capabilities Document (ICD) for Expeditionary Armored Forces (EAF)," MROC Decision Memorandum 07-2008, December 11, 2007.

U.S. Department of the Navy, *Employment of Amphibious Assault Vehicles (AAVs)*, Washington, D.C.: Headquarters U.S. Marine Corps, MCWP 3-13, September 10, 2003.

———, "Medium Tactical Vehicle Replacement (MTVR) Operational Requirements Document (ORD), Change 6," Washington, D.C., January 13, 2004.

U.S. Department of the Navy and U.S. Marine Corps, *Operational Requirements Document for Advanced Amphibious Assault Vehicle (AAAV)*, Acquisition Category I-D, prepared for milestone II decision, No. MOB 22.1, September 13, 2000.

———, "ESOH Integration into System Engineering," presented at the National Defense Industry Association Conference, October 24–27, 2005.

U.S. Government Accountability Office, *Defense Acquisitions: The Expeditionary Fighting Vehicle Encountered Difficulties in Design Demonstration and Faces Future Risks*, Washington, D.C., GAO-06-349, May 1, 2006.

———, *Defense Acquisitions: Assessments of Selected Weapon Programs*, Washington, D.C., GAO-10-388SP, March 2010a.

———, *Defense Acquisitions: Opportunities for the Army to Position Its Ground Force Modernization Efforts for Success*, Washington, D.C., GAO-10-493T, March 10, 2010b.

———, *Defense Management: DoD Needs Better Information and Guidance to More Effectively Manage and Reduce Operating and Support Costs of Major Weapon Systems*, Washington, D.C., GAO-10-717, July 2010c.

———, *Expeditionary Fighting Vehicle (EFV) Program Faces Cost, Schedule, and Performance Risks: Briefing for the Subcommittee on Defense, Committee on Appropriations, House of Representatives*, Washington, D.C., GAO-10-758R, July 2, 2010d.

———, *Defense Acquisitions: Issues to Be Considered as DoD Modernizes Its Fleet of Tactical Wheeled Vehicles*, Washington, D.C., GAO-11-83, November 5, 2010e.

U.S. House of Representatives, Committee on Oversight and Government Reform Majority Staff, *The Expeditionary Fighting Vehicle: Over Budget, Behind Schedule, and Unreliable*, April 29, 2008.

U.S. Joint Chiefs of Staff, "Expeditionary Fighting Vehicle Nunn-McCurdy Certification," Joint Requirements Oversight Council Memorandum 108-07, May 8, 2007.

———, *Department of Defense Dictionary of Military and Associated Terms*, Joint Publication 1-02, Washington, D.C., April 12, 2001, as amended through September 30, 2010.

U.S. Marine Corps, *Concept of Employment (COE) for the Medium Tactical Vehicle Replacement (MTVR)*, undated.

———, "Operational Requirements Document for Advanced Amphibious Assault Vehicle (AAAV) ACAT I-D Prepared for Milestone II Decision," No. MOB 22.1, September 13, 2000.

———, "Operational Requirements Document (ORD) for the Medium Tactical Vehicle Replacement (MTVR) (No. MOB222.4.2, Change 5)," Quantico, Va., May 21, 2001.

———, "Annex B to the Operational Requirements Document (ORD) for the Medium Tactical Vehicle Replacement (MTVR) (No. MOB 211.4.2A)," Quantico, Va., October 16, 2002.

———, "Expeditionary Fighting Vehicle (EFV) Capability Production Document (CPD)," April 13, 2006.

U.S. Marine Corps, Advanced Amphibious Assault Division, "AAAV Acquisition Objective Considerations," briefing, Washington, D.C., undated.

U.S. Marine Corps, Command and Staff College, Marine Corps Combat Development, *Building a Better Mousetrap: The Unnecessary Capability of the EFV EWS 2005*, Quantico, Va.: Marine Corps University, February 8, 2005.

U.S. Marine Corps, Combat Development Command, "Mission Need Statement, Medium Tactical Vehicle Replacement," No. MOB 211.4.2A, March 1992.

———, "Operational Requirements Document (ORD) for the Medium Tactical Vehicle Replacement," No. MOB 211.4.2A, January 1994.

U.S. Marine Corps, Deputy Commandant for Combat Development and Integration, "Expeditionary Fighting Vehicle (EFV) Capability Production Document (CPD); Change 1," September 26, 2007.

———, "Expeditionary Fighting Vehicle (EFV) Capability Production Document (CPD); Change 2," May 20, 2009.

U.S. Marine Corps Operational Test and Evaluation Activity, *Operational Test and Evaluation Manual*, v1.1, October 1, 2009. As of December 28, 2010: http://www.usmc.mil/unit/hqmc/mcotea/PublicDocuments/MCOTEA%20 OTE%20Manual.pdf

U.S. Marine Corps, Expeditionary Fighting Vehicle Program, *Increment: Single Step to Capability Low Rate Initial Production (LRIP), ACAT: I-D*, capability production document prepared for milestone C decision, April 13, 2006.

U.S. Senate, Committee on Armed Services, *National Defense Authorization Act for Fiscal Year 2009*, Report No. 110-335, Washington, D.C.: U.S. Government Printing Office, May 12, 2008.

Viggato, Mike, "Army and Marines Establish the Joint Center for Ground Vehicles," *Accelerate*, Summer 2010, pp. 39–41. As of December 28, 2010: http://tardec.army.mil/Documents/Army%20and%20Marines%20Establish%20 the%20JCGV.pdf

Vane, LTG Michael A., "Support for Congressional Report, NDAA Section 222, Modernization of Ground Combat and Armored Tactical Vehicles," April 26, 2010.

WBB Consulting, *Joint Capabilities Integration and Development System (JCIDS) Documents*, Reston, Va., 2010.

Whaley, D. L., *Advanced Amphibious Assault Vehicle Acquisition Objective Study*, Quantico, Va.: U.S. Marine Corps Combat Development Command, April 2004.